NUMBER THEORY
An Introduction to Proof

NUMBER THEORY
An Introduction to Proof

Charles Vanden Eynden

Associate Professor of Mathematics
Illinois State University

International Textbook Company

An **Intext** Publisher
Scranton, Pennsylvania 18515

ISBN 0-7002-2308-8

3 3001 00572 8893

Preface

This book, designed for one-semester and one-term undergraduate courses, tries to show the *how* and *why* of some of the theorems of number theory. No one would be expected to learn to play the violin merely by attending concerts, yet in our standard courses we show students a sequence of formal proofs and expect them to learn how to make mathematics. That a talented few do just that does not vindicate our method.

The reader is herein invited behind the polished façade of theorem and proof. Consider, for example, the Möbius inversion formula. In the usual treatment the Möbius function is defined, then the formula is stated and proved by the manipulation of summations. In this book examples are first given showing how certain numerical functions F are linked to simpler functions f by the equation $F(n) = \Sigma_{d \mid n} f(d)$. The problem arises of recovering f, given F. Special cases of increasing generality are worked out until it is possible to make a reasonable guess at a formula for f in terms of F. The function μ is defined to simplify the expression of this formula. Finally the formula is proved, care being taken that the summations involved are not perceived only formally.

Of course this takes more time. So be it. The main idea of this book is to develop mathematical maturity and some feeling for the integers. Covering any given theorem was considered secondary.

This is not a book for the mathematically experienced. The shortest path from one theorem to another is not always taken. Proofs were chosen for simplicity, not elegance. The unexpected trick is avoided, even when the alternative is messy. The use of numerical examples is encouraged. The most *natural* development is always sought. The idea is that the author and reader are traveling the road of discovery together. It is hoped that the reader will sometimes be a step ahead.

Answers to the numerical exercises are given in the back of the book. In the same place will be found a set of "extended exercises," suitable for individual assignment, take-home tests, or working in class. Most can be assigned as soon as the fundamental theorem of arithmetic has been covered.

The author would like to acknowledge his debt to many fine number theorists, especially Ivan Niven. He would also like to thank his wife for her help and support.

Charles Vanden Eynden

Normal, Illinois
March, 1970

Contents

Contents

Introduction

One of the most frequently used words in this book is "we." Turn to almost any page and you will see "*we* know that . . . ," "thus *we* see . . . ," "*we* must show that . . . ," etc. This is not the editorial "we" or the royal "we"; neither does it indicate more than one author. In this book "we" means *the author and the reader*—YOU and I.

This book is not meant to be read passively. The reader must *participate*. Nobody would expect to learn to swim or to play the harmonica *only* from books. Neither should anybody expect to learn mathematics by just reading about it. Mathematics must be *practiced*.

Of course one can learn *facts* about swimming or playing the harmonica from a book. This book also contains facts—facts about number theory. Facts are important. But imparting facts is not the main purpose of this book. The main goals of this book are *technique* and *understanding*. Let me explain why these are more important than facts.

First of all, mathematical facts can be recovered, given technique. Suppose, for example, you need the formula for sin $(x + y)$. You don't remember it, and there is no place handy to look it up. Missing: one trigonometric fact. In this situation anyone with a modicum of mathematical competence would merely draw a little picture and figure out the formula. Technique to the rescue.

Now let me give an example to illustrate the distinction I make between facts and understanding, and how the latter helps in the acquisition of the former. You are a baseball fan, watching the World Series on television. A friend walks in and asks what happened in the last half-inning. You tell him the names of the batters in order and what each did at the plate, whether he struck out, walked, flied out (and to what field), got a hit, and so on. You describe the situation on the bases, the number of outs, and the score at each stage of the inning. You tell about any stolen bases, unusual fielding plays, pinch hitters, or pitching changes. I could go on and on about the amount you remember about what you have just seen—*without any conscious effort to remember anything*.

I claim *understanding* is the key. You perceive the events you watched as an interconnected whole. You can do this because you understand the mechanics of baseball. Suppose, for example, that the first man up struck out and the second man walked. Wouldn't it be a natural mistake to get these two events

1

reversed in time? Not at all. The situation would be completely different. A man on first with one out is much different than a man on first with no outs. Player A on first represents a different situation than player B.

Suppose the game had been cricket instead of baseball. Let us assume you know as little about cricket as the average American. How well could you reconstruct what you saw? See what I mean?

The baseball example show how remembering comes easy when you have understanding. The *more* understanding, the better off you are. You leave the room. When you come back you ask your Aunt Edna (who wanted to watch "The Secret Storm") what has happened. "The first three men with red hats struck out and the fourth hit a home run." She could have the world's worst memory and still not make that mistake—if she knew a little more about the game. On the other hand, someone *more* involved in baseball than you would likely remember even more about the inning. The manager of the team in the field could probably recall every pitch.

Something else is involved here. If you are interested in a baseball game, you don't watch it in a passive way, even though you may appear passive to Aunt Edna. ("Why aren't you mowing the lawn?") Your *mind* is active, thinking not only about what you have already seen, but also about what you *might* see. Even when there is no action in the ball park you are at work, considering the possibilities. If the man at the plate should get on base, who is up next? Should the pitcher be taken out? Did the manager make a mistake in not pinch-hitting for him last inning? (Of course the manager's mind is even more active; he may be thinking ahead to tomorrow's game.)

It is in this active way that this book should be read. You should second-guess me just as you would the baseball manager. What am I going to do next? Am I proving something in the easiest way? Going down a blind alley? Is that statement right? Does the relation hold for zero, for negative numbers? In fact, reading this book properly requires more than just mental activity. One should no more sit down with any mathematics book without a pencil and paper close at hand than start driving up Pike's Peak in January without chains.

What are the paper and pencil for? To pull you through the tough places, when you don't quite understand a step or statement. To work out numerical examples of definitions and theorems. To do the exercises (more about these shortly). And to look for counterexamples where you think I've fouled it up. The essence of the proper attitude for reading this book is extreme skepticism. Remember that in mathematics *nothing is right just because somebody says it is*. Everything must be proved. Imagine you are the manager of a hotel hosting a convention of swindlers. Don't let me put anything over on you.

The exercises form an integral part of this book. They are not collected at the end of chapters but are scattered throughout the text. The best time to do an exercise is when it appears. A few exercises are marked with a star; these contain results used in the text. No starred exercise should be skipped (otherwise I may be putting something over on you).

Some exercises are computational (answers to these appear in the back of the book), but most call for proofs. Since constructing proofs is what this book is all about, it may be worthwhile to put in a few words about the subject here. First a pessimistic note. Finding a proof is invariably a sometimes thing. A good student might plan to do an assigned set of calculus problems between 9 and 10 P.M. on Sunday. No one can say in advance, however, how long it will take person A to do proof B, or even if he will ever get it. It may depend on some simple wrinkle that just never pops into his head. Thus after spending a reasonable amount of time on a problem without success it is probably a good idea to go on to something else. Many mathematicians report solving their most difficult problems with the aid of the unconscious mind. Sleeping on a problem often helps. Of course the problem must be thoroughly understood to begin with; the unconscious needs something to work with.

Proving something can be broken into two phases. The first is discovering the proof; the second is writing it up in a formal way. One should no more expect to combine these two steps into one than one should expect to write a good novel in a single draft. For one thing, the final write-up often reveals a hole in what seemed to be an airtight argument. Another reason for separating these two steps is that the rigorous standards of logic that must be applied to any proof in the end inhibit the free, anything-goes attitude that fosters discovery. It is said that the best jazz performances don't happen in recording studios, where the pressure of so many takes induces a conservatism in the musicians. In the same way, the thought that someone might read what you are writing may cause your mind to reject the "crazy" idea that contains the germ of a proof.

In this book I try to show not only what proofs are but how they are made. Our route to a proof may be devious. We may go up dead-end streets, even make mistakes. The final use of the word "theorem" may be justified by a pyramid of arguments stretching back for pages. Sometimes at this stage I will reprise the proof in a formal way, eliminating what was irrelevant or superseded by more general arguments and arranging the steps in the most efficient way. (This is the sort of thing you will see following the word "Proof" in most mathematics books.) More often I omit this formal write-up. You, the active reader, should supply it.

There are many sets of True-False exercises in this book. The idea is to prove those statements that are true and disprove the others. The disproofs ordinarily take the form of a counterexample. The more concrete and definite the counterexample is, the better. For example, consider the statement

$$\text{If } a \geqslant b, \text{ then } ac \geqslant bc.$$

This could be disposed of by

False. Let $a = 2$, $b = 1$, and $c = -1$. Then $a = 2 \geqslant 1 = b$, but $ac = -2$ and $bc = -1$. Thus $ac \geqslant bc$ is false.

This is superior to

False. If $a \geqslant b$ and c is negative, then $ac < bc$.

The second disproof is inferior in two respects. First, it is more general than it has to be, invoking a theorem about inequalities when all that is really needed is that $-2 < -1$. Second, it is *itself* false. What if $a = b = 0$ and $c = -1$? Then $a \geqslant b$ and c is negative, yet $ac < bc$ is false. True enough, the second disproof is valid except in the special case that $a = b$. But that is enough to invalidate it. Mathematics is a very *picky* subject.

It should be mentioned that a purely numerical counterexample does not always suffice to disprove a statement. Consider

There exists x such that $x > y$ for all y.

It is not good enough to say

False. Let $x = 100$. Then if $y = 200, x > y$ is false.

After all, nobody said x was 100. It must be shown that *no x* satisfies the statement.

False. Let any x be given. Set $y = x + 1$. Then $x > y$ is false.

The above is a satisfactory disproof.

Sometimes when you are trying to prove something and are stumped, it helps to look for a counterexample. If you can see *why* you can't find one, you are on your way to a proof.

It is a ground rule that in proving an exercise any previous exercise or result of the text may be assumed unless the contrary is expressly stated.

Anyone who has read this far in a number theory book deserves some word on what our subject is about. No completely satisfactory definition of number theory has ever been given. Roughly, it is *the mathematical treatment of questions related to the integers*. These are the whole numbers, positive, negative, and zero.

chapter 1

Divisibility

The sum, difference, and product of two integers is again an integer, but the quotient need not be. This chapter is an investigation of the interesting structure arising from this seeming defect in the system of whole numbers, and includes a closer look at many of the familiar ideas of arithmetic.

1 THE LEAST COMMON MULTIPLE AND GREATEST COMMON DIVISOR

(1) A person asked to add $^3/_4$ and $^1/_6$ will probably write something like

$$^3/_4 + ^1/_6 = ^9/_{12} + ^2/_{12} = ^{11}/_{12}.$$

The 4ths and 6th are converted to 12ths, 12 being the "least common multiple" of 4 and 6. We review the definition of this concept, which the reader undoubtably first learned in grade school.

(2) **Definition.** Suppose a and b are integers. If a/b is also an integer we say that a is a *multiple* of b, or that b is a *divisor* of a, or that b *divides a*. We express this symbolically by $b|a$. If it is false that $b|a$ we write $b\nmid a$.

(3) **Examples.** 6 is a multiple of 3; 2 divides 6; 4 is a divisor of -12; $-1|5$; $7|0$; $6\nmid 9$; $0\nmid 7$.

(4) **Definition.** The *least common multiple* of a and b is the smallest among those positive numbers which are multiples of both a and b. It is denoted by $[a,b]$.

(5) **Examples.** $[6,9] = 18$; $[-4,6] = 12$; $[-3,-5] = 15$.

(6) *Note.* As long as neither a nor b is 0 at least one positive common multiple exists, namely $|ab|$. Thus $[a,b]$ exists and $[a,b] \leqslant |ab|$.

(7) If the reader can add fractions he presumably has some way of computing the least common multiple of a pair of positive integers. One method is to start listing the positive multiples of each number and then compare lists. Thus in order to compute $[6,4]$ we might write the multiples of 6: 6, 12, 18, 24, . . . ,

and of 4: 4, 8, 12, 16, . . ., and then note that 12 is the smallest member of both lists. Although each number has infinitely many multiples, we need only take our lists so far, since we know $[6,4] \leqslant 6 \cdot 4 = 24$.

(8)

a	b	$[a,b]$	ab
6	3	6	18
6	4	12	24
6	5	30	30
6	6	6	36
6	7	42	42

The reader is asked to stop at the end of this sentence and inspect the above table, with the aim of discovering what distinguishes those pairs of positive integers the least common multiple of which is strictly less than their product.

(9) Perhaps the reader has noticed that if the positive integers a and b have a common *divisor* greater than 1, then $[a,b] < ab$. For example 2 divides both 6 and 10, and $[6,10] = 30 < 6 \cdot 10$. Although $60 = 6 \cdot 10$ is *a* multiple of both $6 = 2 \cdot 3$ and $10 = 2 \cdot 5$, 60 includes the factor 2 twice when once would suffice. Since $3 \cdot 5 = 15$ is a multiple of both 3 and 5, $2 \cdot 15$ will be a multiple of both $2 \cdot 3 = 6$ and $2 \cdot 5 = 10$.

(10) **Proposition.** If the positive integers a and b have a common divisor greater than 1, then $[a,b] < ab$.

   Proof. Let the common divisor be d. We claim that ab/d is a common multiple of a and b. This is true since $(ab/d)/a = b/d$ and $(ab/d)/b = a/d$ are both integers. Then $[a,b] \leqslant ab/d < ab$, since $[a,b]$ is the *least* common multiple and $d > 1$.

(11) We see we have really proved more than the statement of (10), namely:

(12) **Proposition.** If a, b, and d are positive integers and d divides both a and b, then $[a,b] \leqslant ab/d$.

(13) The preceding proposition says the most when ab/d is smallest; i.e., when d is biggest.

(14) **Definition.** The *greatest common divisor* of the integers a and b is the largest among those integers which divide both a and b. It is denoted by (a,b).

(15) **Examples.** $(6,10) = 2; (-5, -3) = 1; (7,0) = 7$.

(16) *Note.* At least one common divisor of a and b always exists, namely 1. If $a \neq 0$ the common divisors of a and b cannot exceed $|a|$. Thus (a,b) exists as long as not both a and b are 0.

(17) **Proposition.** If a and b are positive integers, then $[a,b] \leqslant ab/(a,b)$.

   Proof. See (12).

(18) The definition of the greatest common divisor is motivated by our trying

to squeeze the strongest possible statement from the argument of (10). To see how near the truth we have come we extend our previous table to include (a,b) and $ab/(a,b)$.

a	b	$[a,b]$	ab	(a,b)	$ab/(a,b)$
6	3	6	18	3	6
6	4	12	24	2	12
6	5	30	30	1	30
6	6	6	36	6	6
6	7	42	42	1	42

(19) The reader is asked to examine the third and last columns of the above table and to make a conjecture.

(20) *Conjecture.* If a and b are positive integers,

$$[a,b] = ab/(a,b).$$

(21) Of course the fact that the above equation is true in five instances does not prove its validity in general. It does encourage us to attempt a proof. This points up an advantage number theorists have: a concrete subject (the integers), in which theorems may be suggested by numerical data.

(22) We want to show $[a,b] = ab/(a,b)$ and we already know $[a,b] \leqslant ab/(a,b)$. Thus to confirm our conjecture it suffices to show that $[a,b] \geqslant ab/(a,b)$. Of course the proof must depend somehow on the special properties of $[a,b]$ and (a,b). We have already used [in (10)] the fact that $[a,b]$ is the *least* common multiple of a and b. Nowhere, however, have we yet used the fact that (a,b) is their *greatest* common divisor, only that it is *some* divisor of both. Since $[a,b]$ and (a,b) are both positive, the inequality we are after is equivalent to $(a,b) \geqslant ab/[a,b]$. How might we show this?

(23) If we knew that $ab/[a,b]$ was a divisor of both a and b, the inequality $(a,b) \geqslant ab/[a,b]$ would follow from the definition of (a,b). To show that $ab/[a,b]$ is a divisor of a, we must prove that $a/(ab/[a,b])$ is an integer. But $a/(ab/[a,b]) = [a,b]/b$, which is an integer since $[a,b]$ is a multiple of b. Likewise $b/(ab/[a,b]) = [a,b]/a$, which is an integer. Thus $ab/[a,b]$ also divides b; $(a,b) \geqslant ab/[a,b]$; and our conjecture follows.

(24) There is something wrong with the argument given above. If he has not already noticed the error, the reader is invited to try to find it before going on.

(25) The argument in (23) is perfectly valid, so long as we know that $ab/[a,b]$ *is an integer*. We have not defined "x divides y" except for x and y integers.

(26) We proceed to show that $ab/[a,b]$ is an integer. We can certainly write the real number $ab/[a,b]$ as $k + \epsilon$, where k is an integer and $0 \leqslant \epsilon < 1$. For example, $\pi = 3 + .14159\ldots$, $-9/5 = -2 + 1/5$, $13 = 13 + 0$. Multiplying by $[a,b]$, we get $ab = [a,b]k + [a,b]\epsilon$. Let $[a,b]\epsilon = r$. Then $r = ab - [a,b]k$. We see r is an integer. Also $0 \leqslant r < [a,b]$.

Now $[a,b]$ is a multiple of a, so $[a,b]/a$ is an integer, say l. Then $r/a = ab/a - [a,b]k/a = b - lk$, also an integer. Therefore $a|r$. In the same fashion it can be shown that $b|r$. Thus r is a common multiple of a and b. Since $[a,b]$ is the *least* (positive) common multiple of a and b, and $0 \leqslant r < [a,b]$, we must have $r = 0$. We see that $ab/[a,b]$ is the integer k. Thus we have

(27) **THEOREM**. If a and b are positive integers, then $[a,b] = ab/(a,b)$.

(28) *Note*. In the preceding paragraphs the author has tried to indicate how a mathematician might have discovered the above theorem experimentally and then found a proof for it. If he decided to put his result in a paper or a text-book, however, he would probably erase his steps and present it in a totally different way. He would start by defining the least common multiple and greatest common divisor, and then immediately state the theorem. Then would come the proof, first showing that $[a,b] \leqslant ab/(a,b)$ by the argument of (10), then reversing the inequality by means of (26) and (23). Such a presentation has the advantages of compactness and adherence to tradition, and is not (usually) meant to be obscure. It is preferred by the experienced mathematical reader. It offers no particular guidance, however, to the beginning theorem inventor and prover, for whom the definitions seem to come from left field and the theorem as a bolt from the blue. The author proposes to start in an (admittedly slow) "organic" style which approximates how mathematics is actually made, coming around to a more usual presentation as the reader develops the ability to fill in for himself the dismantled framework of a discovery.

(29) Incidentally, the astute reader may have noticed that the definition of greatest common divisor given above was tailored exactly so as to make the theorem true. This practice is common in mathematics and is in no way considered shady. It will turn out that the idea of the greatest common divisor arises in many contexts and fully justifies a special name and symbol.

(30) **Exercise**. Make a table like that in (18) for a running from 6 through 10 and $b = -12$.

(31) **True-False**. Below a, b, and c are assumed to be arbitrary nonzero integers.
    (a) If $ab > 0$, then $[a,b] \leqslant ab$.
    (b) If $c|a$ and $c|b$, then $[a,b] \leqslant ab/c$.
    (c) If $(a,b) = 1$ and $(a,c) = 1$, then $(b,c) = 1$.
    (d) $[a,b](a,b) = |ab|$.
    (e) If $b|c$, then $(a,b) \leqslant (a,c)$.
    (f) If $b|c$, then $[a,b] \leqslant [a,c]$.
    (g) If $a|b$ and $b|c$, then $a|c$.
    (h) $(ac, bc) = c(a,b)$.
    (i) $(ac,bc) = |c|(a,b)$.

2 THE DIVISION ALGORITHM

(32) At this point the reader is invited to reread (26), since the argument used there suggests a number of generalizations which we will proceed to develop. He will notice that we started by expressing ab as a multiple of $[a,b]$ plus an integer r, where $0 \leqslant r < [a,b]$. All we needed to get this far was that $[a,b]$ was positive. This is really a familiar arithmetic fact. For example, one knows without doing the actual computation that if 592 is divided by 7 the answer is some integer q plus $r/7$, where $r = 0, 1, 2, 3, 4, 5$, or 6. Thus (multiplying by 7) $592 = 7q + r, 0 \leqslant r < 7$.

(33) **The Division Algorithm.** Suppose a and b are integers, $a > 0$. Then there exist integers q and r, $0 \leqslant r < a$, such that $b = aq + r$.

Proof. Suppose the real number b/a is $q + \epsilon$, where q is an integer and $0 \leqslant \epsilon < 1$. Then $b = a(q + \epsilon) = aq + a\epsilon$. We easily see that $a\epsilon$ is an integer and $0 \leqslant a\epsilon < a$. Set $r = a\epsilon$.

(34) In (26) a simple argument was used to show that since a divided both ab and $[a,b]$ it followed that a divided $r = ab - [a,b] k$. The following theorem is a generalization of this idea. Its proof is left to the reader.

(35) **THEOREM.** If $d | a_i, i = 1, 2, \ldots, n$, and if $x_1, x_2, \ldots, x_n$ are any integers, then

$$d | x_1 a_1 + x_2 a_2 + \ldots + x_n a_n.$$

(36) Still more can be mined from (26). Recall that the idea was to show that $ab/[a,b]$ was an integer. Looking back over the proof we see that the only property of ab that was needed was that it was a multiple of both a and b. In other words,

(37) **THEOREM.** If m is any common multiple of a and b, then $[a,b] | m$.

Proof. By the Division Algorithm there exist integers q and r, $0 \leqslant r < [a,b]$, such that $m = [a,b] q + r$. Since $a | m$ and $a | [a,b]$, (35) shows $a | r$. Likewise $b | r$. Again r is a common multiple of a and b; by the minimality of $[a,b]$ we must have $r = 0$. Thus $[a,b] | m$.

(38) The reader may have noticed a certain duality between the concepts of least common multiple and greatest common divisor. Replacing "least" with "greatest" and "multiple" with "divisor" changes the definition of the former into that of the latter. New (possible) theorems can be created in a like way. For example (37) "dualizes" into

(39) **THEOREM.** If d is any common divisor of a and b, not both of which are zero, then $d | (a,b)$.

Proof. Of course this requires a proof in its own right. The desired result is so much like (37), however, that a promising approach would be to use (37) and the relation $(a,b) = ab/[a,b]$ from (27). We want to show that if $d|a$ and $d|b$, then $d|ab/[a,b]$. Saying $d|ab/[a,b]$ is equivalent to saying $ab/d[a,b]$ is an integer, or, again, that $[a,b]$ divides ab/d. (What we are trying to do is work the problem around so that (37) can be applied.) But (37) says that $[a,b]$ divides any common multiple of a and b. We see that $ab/d = (a/d)b = a(b/d)$ fills the bill.

It should be pointed out that the above proof is valid only for a and b both positive, since this was assumed in (27). This case implies the theorem for arbitrary nonzero a and b however, since if $d|a$ and $d|b$, then d divides both $|a|$ and $|b|$. Thus by the above $d|(|a|, |b|) = (a,b)$. The theorem is easy for one of a or b zero.

(40) **Exercise.** Find q and r as guaranteed by the Division Algorithm for $a = 13$, $b = 380$; for $a = 380, b = 13$; for $a = 13, b = -380$.

(41) **Exercise.** Prove that the q and r of the Division Algorithm are unique (i.e., prove that if $aq + r = aq' + r'$, with $0 \leqslant r < a$ and $0 \leqslant r' < a$, then $q = q'$ and $r = r'$.)

(42) **True-False.** Below a, b, c, and d are arbitrary integers, with $a > 0$, and c and d nonzero.
   (a) There exist integers q and r, $0 \leqslant r < c$, such that $b = cq + r$.
   (b) There exist integers q and r, $0 \leqslant r < |c|$, such that $b = cq + r$.
   (c) There exist integers q and r, $|r| \leqslant a/2$, such that $b = aq + r$.
   (d) There exist integers q and r, $|r| < a/2$, such that $b = aq + r$.
   (e) The set of common multiples of c and d is the set of multiples of $[c,d]$.
   (f) The set of common divisors of b and c is the set of divisors of (b,c).
   (g) If b is a divisor of c, and $b > (c,d)$, then b is not a divisor of d.
   (h) If b is a multiple of c, and $b < [c,d]$, then b is not a multiple of d.

3 THE EUCLIDEAN ALGORITHM

(43) Consider the equation $b = aq + r$. By (35) any common divisor of a and r also divides b; in particular $(a,r)|b$. But $r = b - aq$, so the same argument tells us that $(a,b)|r$. Since the "common part" of a and r divides b and vice versa, we might guess that $(a,b) = (a,r)$. This is easily confirmed. By definition, $(a,r)|a$ and we just saw $(a,r)|b$. Thus $(a,r) \leqslant (a,b)$, by the definition of the latter. (Actually $(a,r)|(a,b)$ by (39).) In the same way (a,b) is a common divisor of a and r, so $(a,b) \leqslant (a,r)$. We see $(a,b) = (a,r)$.

(44) The fact developed above can be used to simplify the calculation of the greatest common divisor of two integers. We may assume the integers are posi-

tive, since $(x,y) = (|x|,|y|)$. Call the smaller a and the larger b. By the Division Algorithm we can write $b = qa + r, 0 \leqslant r < a$. By (43) we know that $(b,a) = (a,r)$. But $a < b$ and $r < a$, so we may expect (a,r) to be easier to calculate than (b,a). Suppose, for example, we want $(504,123)$. We divide 123 into 504 to get 4, with remainder 12. Thus $504 = 4 \cdot 123 + 12$. Then $(504,123) = (123,12)$. Of course we needn't stop here. In the same way we write $123 = 10 \cdot 12 + 3$, and conclude $(123,12) = (12,3)$. Although $(12,3)$ is easily seen to be 3, let us see how far we can carry this process. We have $12 = 4 \cdot 3 + 0$, and so $(12,3) = (3,0)$. Of course we stop here, since 0 cannot be a divisor.

Let us rewrite our equations, using arrows to emphasize the pattern.

$$504 = 4 \cdot 123 + 12, \quad 0 \leqslant 12 < 123,$$
$$123 = 10 \cdot 12 + 3, \quad 0 \leqslant 3 < 12,$$
$$12 = 4 \cdot 3 + 0, \quad 0 \leqslant 0 < 3.$$

(45) **The Euclidean Algorithm.** Suppose a and b are integers, $a > 0$. Apply the Division Algorithm repeatedly as follows:

$$b = q_1 a + r_1, \quad 0 \leqslant r_1 < a,$$
$$a = q_2 r_1 + r_2, \quad 0 \leqslant r_2 < r_1,$$
$$r_1 = q_3 r_2 + r_3, \quad 0 \leqslant r_3 < r_2,$$

$$\ldots$$

$$r_{n-2} = q_n r_{n-1} + r_n, \quad 0 \leqslant r_n < r_{n-1},$$
$$r_{n-1} = q_{n+1} r_n + 0, \quad 0 \leqslant 0 < r_n.$$

Suppose r_n is the last nonzero remainder generated in this way. [In the example just given $n = 2$ and $r_n = 3$.] Then $(a,b) = r_n$.

Proof. The process must terminate, since the remainders form a strictly decreasing sequence of nonnegative integers. By (43) we see $(a,b) = (a,r_1) = (r_1,r_2) = \ldots = (r_{n-1},r_n) = (r_n,0)$. But $(r_n,0) = r_n$.

4 LINEAR COMBINATIONS

(46) **Definition.** We say that k is a *linear combination* of the integers $a_1, a_2, \ldots, a_n$ in case there exist integers $x_1, x_2, \ldots, x_n$ such that $k = x_1 a_1 + x_2 a_2 + \ldots + x_n a_n$.

(47) **Exercise***.[1] Prove that if k is a linear combination of a_1 and a_2, and if a_1 and a_2 are each linear combination of b_1 and b_2, then k is a linear combination of b_1 and b_2.

---

[1] Starred exercises are used later in the text and so should invariably be worked.

(48) Since, in the notation of (45), $r_1 = b - q_1 a$, we see r_1 is a linear combination of a and b. Likewise $r_2 = a - q_2 r_1$ is a linear combination of a and r_1. But then r_2 is also a linear combination of a and b [see (47)]. This argument can be continued all the way to $r_n = (a,b)$.

(49) **THEOREM.** For any integers a and b, not both 0, there exist integers x and y such that $(a,b) = xa + yb$.

(50) The integers x and y can be found explicitly from the equations of the Euclidean Algorithm by starting with r_n and successively eliminating r_{n-1} through r_1. For example, from the equations of (44),

$$(123,504) = 3 = 123 - 10(12)$$
$$= 123 - 10(504 - 4 \cdot 123)$$
$$= 41(123) - 10(504).$$

Here $x = 41$ and $y = -10$.

(51) **Example.** We calculate $(302,1041)$.

$$1041 = 3 \cdot 302 + 135$$
$$302 = 2 \cdot 135 + 32$$
$$135 = 4 \cdot 32 + 7$$
$$32 = 4 \cdot 7 + 4$$
$$7 = 1 \cdot 4 + 3$$
$$4 = 1 \cdot 3 + 1.$$

Clearly 1 is the last nonzero remainder, so $(302,1041) = 1$. Now

$$1 = 4 - 1 \cdot 3$$

$$= 4 - 1(7 - 1 \cdot 4) \qquad\qquad = 2(4) - 7$$
$$= 2(32 - 4 \cdot 7) - 7 \qquad\qquad = 2(32) - 9(7)$$
$$= 2(32) - 9(135 - 4 \cdot 32) \qquad = 38(32) - 9(135)$$
$$= 38(302 - 2 \cdot 135) - 9(135) \qquad = 38(302) - 85(135)$$
$$= 38(302) - 85(1041 - 3 \cdot 302) = 293(302) - 85(1041).$$

Thus $x = 293$ and $y = -85$.

(52) **Exercise.** Calculate (a,b) by means of the Euclidean Algorithm and then find integers x and y such that $(a,b) = xa + yb$ for $a = 287$, $b = 567; a = 1597, b = 987; a = -101, b = -1001$.

(53) **Exercise.** Suppose that S is a nonempty set of integers having the property that whenever a and b are in S, then $a + b$ and $a - b$ are in

S. Show that there exists a nonnegative integer c in S such that S is exactly the set of all multiples of c.

(54) **Exercise.** Construct an alternate proof of (49) along the following lines. Let c be the least positive linear combination of a and b. Divide a by c and show that the remainder is also a linear combination of a and b. Conclude that $c|a$. Show similarly that $c|b$. Thus $c \leqslant (a,b)$ and (49) follows.

(55) **Exercise.** Determine all integers x such that there exists an integer y such that $2x + 3y = 1$.

(56) Suppose $(a,b) = xa + yb$. Then $c(a,b) = (cx)a + (cy)b$ for any integer c. We see that any multiple of (a,b) is also a linear combination of a and b. The converse follows from (35).

(57) **THEOREM.** An integer k is a linear combination of a and b if and only if $(a,b)|k$.

(58) *Note.* The preceding theorem may be thought of as solving a "representation" problem, in this case the problem of which integers can be represented in the form $xa + yb$, where a and b are fixed and x and y arbitrary integers. Such problems are common in number theory. Another example is the determination of all integers expressible in the form $x^2 + y^2$. One such integer is $5 = 1^2 + 2^2$; one that isn't is 6. Of special interest are forms representing all integers (or, sometimes, all positive integers). For instance, it can be shown that every positive integer is the sum of 4 squares, while some are not the sum of 3 squares. (The reader is invited to find the smallest example of the latter.) Our theorem shows that $xa + yb$ represents all integers if and only if $(a,b) = 1$.

(59) **THEOREM.** There exist integers x and y such that $xa + yb = 1$ if and only if $(a,b) = 1$.

(60) **Definition.** If $(a,b) = 1$ we say that a and b are *relatively prime*.

5 CONGRUENCES

(61) We continue our study of (a,b). To simplify things let us hold a fixed. The following table lists the values of $(6,b)$ for $b = 1, 2, \ldots, 20$.

b	$(6,b)$	b	$(6,b)$	b	$(6,b)$	b	$(6,b)$
1	1	6	6	11	1	16	2
2	2	7	1	12	6	17	1
3	3	8	2	13	1	18	6
4	2	9	3	14	2	19	1
5	1	10	2	15	3	20	2

(62) **Exercise**. Construct tables for $(4,b)$ and $(5,b)$ as b runs from 1 through 20.

(63) Examination of the table of (61) reveals a repetition of the cycle $1,2,3,$ $2,1,6$. It appears that the numbers $1, 7, 13, 19, \ldots$ all have the same greatest common divisor with 6; likewise $2, 8, 14, 20, \ldots$; etc. The common characteristic within these groups of numbers is that their members all differ from one another by multiples of 6.

(64) **Definition**. Let a be a positive integer. We say that the integers b and b' are *congruent modulo a* and write $b \equiv b' \pmod{a}$ in case $a|b - b'$.

(65) **THEOREM**. Two numbers are congruent modulo a if and only if they produce the same remainder r when divided by a according to the Division Algorithm.

Proof. Suppose
$$b = qa + r, \ 0 \leqslant r < a,$$
and
$$b' = q'a + r', \ 0 \leqslant r' < a.$$
If $b \equiv b' \pmod{a}$, then $a|b - b' - (q - q')a = r - r'$. But $-a < r - r' < a$ and so $r - r' = 0$. Conversely, if $r = r'$, then $b - b' = (q - q')a$.

(66) **THEOREM**. If $b \equiv b' \pmod{a}$, then $(a,b) = (a,b')$.

Proof. Let $b - b' = ka$. Then $b = ka + b'$. Exactly as in (43) $(b,a) = (a,b')$.

6 PRIME NUMBERS

(67) The last theorem shows that in order to investigate (a,b) for fixed $a > 0$ it suffices to consider only b between 1 and a. It is easy to see that the set of values of (a,b) is exactly the set of positive divisors of a. For $a = 6$ these are $1, 2, 3,$ and 6. The larger number 7 has only two positive divisors, 1 and 7.

(68) **Definition**. A positive integer is said to be *prime* in case it has exactly two positive divisors. A positive number having more than two positive divisors is said to be *composite*.

(69) *Note*. According to the above 1 is neither prime nor composite.

(70) **Examples**. The primes less than 20 are $2, 3, 5, 7, 11, 13, 17,$ and 19.

(71) Experimentation quickly convinces one that any number not itself prime, such as 30, can be broken down into prime factors, as $2 \cdot 3 \cdot 5$. This can be proved in a straightforward way.

(72) **THEOREM**. Suppose $n > 1$ is an integer. Then $n = p_1 p_2 \ldots p_t$ where the p's are all prime.

Proof. If n is itself prime we are done. Otherwise n must have a factor, say d, other than itself and 1. Let $n = dd'$. Clearly $1 < d' < n$ also. We now apply the same argument to d and d' as we did to n. This procedure must end since the factors grow smaller at each step. But it can stop only when each factor has no positive divisors other than 1 and itself; i.e., when each factor is prime.

7 THE τ FUNCTION

(73) **Definition.** We define $\tau(n)$ to be the number of positive divisors of n.

(74)

n	Positive divisors of n	$\tau(n)$
1	1	1
2	1,2	2
3	1,3	2
4	1,2,4	3
5	1,5	2
6	1,2,3,6	4
7	1,7	2
8	1,2,4,8	4
9	1,3,9	3
10	1,2,5,10	4

It would be nice if we had some way to compute $\tau(n)$ beside the brute-force method of writing out all the divisors. Finding $\tau(1,000,000)$ this way would be tedious. We might try to approach a formula for $\tau(n)$ by starting with special values of n. We see from the above table, for example, that if p is prime, $\tau(p) = 2$. This is a case of biting off less than we can chew, however, since this is merely our definition of "prime." A more nourishing observation is that if p_1 and p_2 are distinct primes, then $\tau(p_1 p_2) = 4$. This certainly holds for $2 \cdot 3$ and $2 \cdot 5$, and $\tau(3 \cdot 5)$ is easily seen to be 4 also. Proving even this seemingly simple result offers an immediate difficulty, however. Of course $p_1 p_2$ has divisors $1, p_1, p_2$, and $p_1 p_2$; the trouble comes in proving that these are its *only* positive divisors. Although it is transparent that 1, 3, 5, and 15 are the only positive divisors of 15, it is somewhat less clear that $211 \cdot 227$ has only 4 positive factors (211 and 227 are primes).

(75) **Exercise*.** Suppose that p_1 and p_2 are primes, and that p_1 and p_2 are the only primes dividing $p_1 p_2$. Show that $\tau(p_1 p_2) = 4$.

(76) The exercise shows that it suffices to prove that if p is a prime different from p_1 and p_2, then $p \nmid p_1 p_2$. Since we have only one theorem about primes, let us try to restate our problem in terms of concepts we know more about. The only positive divisors of p are 1 and p, so $(p,b) = 1$ or p no matter what b is. If $p \mid b$, $(p,b) = p$; if $p \nmid b$, $(p,b) = 1$. Thus $p \nmid b$ if and only if $(p,b) = 1$. We can use this to translate "if $p \neq p_1$ and $p \neq p_2$, then $p \nmid p_1 p_2$" into "if $p \neq p_1$ and $p \neq p_2$, then $(p,p_1 p_2) = 1$." The advantage of this formulation is that we already know a good deal about the greatest common divisor function. Since

distinct primes are clearly relatively prime (and equal primes not), we can push our translation all the way to

(77) **Proposition.** If $(p,p_1) = 1$ and $(p,p_2) = 1$, then $(p,p_1p_2) = 1$.

(78) This is something we might expect to be true whether the numbers involved were prime or not. Inspection of the table of (61), for example, suggests that for any integers a, b, and b' such that $(a,b) = (a,b') = 1$, we have $(a,bb') = 1$. Another way to say this is that the set of numbers relatively prime to a is closed under multiplication. Thus $(6,5) = (6,7) = 1$, and also $(6,35) = 1$.

We can get an algebraic characterization of the numbers relatively prime to a by means of (59). That is, $(a,b) = 1$ if and only if there exist integers x and y such that $xa + yb = 1$, or $b = (1 - xa)/y$. Thus the numbers relatively prime to a are exactly the integers of the form $b = (1 - xa)/y$. But suppose $b' = (1 - x'a)/y'$. Then $bb' = (1 - xa)(1 - x'a)/yy'$, $= (1 - xa - x'a + xx'a^2)/yy' = (1 - (x + x' - xx'a)a)/yy'$. Thus bb' is again of this form and $(a,bb') = 1$.

We will now recap this argument in a more formal way. First we must take care of the detail that in writing $b = (1 - xa)/y$ we may be dividing by 0. This is a small point, but there is no minimum size for the gaps in reasoning that invalidate a proof. This one is easily chinked.

(79) **Proposition.** $(a,b) = 1$ if and only if there exist integers x and y such that $b = (1 - xa)/y$.

Proof. If $b = (1 - xa)/y$, then $xa + yb = 1$ and so $(a,b) = 1$ by (59). For the converse, suppose $(a,b) = 1$. Then there exist x and y such that $xa + yb = 1$.

Case I. $y \neq 0$. Then $b = (1 - xa)/y$.

Case II. $y = 0$. Then $xa = 1$ so $a = \pm 1$. Let $y' = 1$ and $x' = (1 - b)/a$. Then $(1 - x'a)/y' = 1 - ((1 - b)/a)a = b$.

(80) **THEOREM.** Suppose $(a,b) = (a,b') = 1$. Then $(a,bb') = 1$.

Proof. By (79) there exist integers x, y, x', and y' such that $b = (1 - xa)/y$ and $b' = (1 - x'a)/y'$. Then $bb' = (1 - (x + x' - xx'a)a)/yy'$ so $(a,bb') = 1$, again by (79).

8 UNIQUE FACTORIZATION

(81) In trying to prove $\tau(p_1p_2) = 4$, we have come up with a much more general and important result. Recalling that for p prime, $(p,b) = 1$ if and only if $p \nmid b$, we can translate (80) back to "if $p \nmid b$ and $p \nmid b'$, then $p \nmid bb'$," or, what is the same,

(82) **THEOREM.** If the prime p divides bb', then $p | b$ or $p | b'$.

(83) Combining this theorem with (75) we see that the only ways to factor $p_1 p_2$ into primes are as $p_1 p_2$ and $p_2 p_1$. A much more general result holds.

(84) **Proposition.** Suppose $p_1 p_2 \ldots p_s = q_1 q_2 \ldots q_t$, where the p's and q's are all primes. Then $s = t$, and the q's are just the p's back again (perhaps in some other order).

Proof. We claim p_1 is one of the q's. Since $p_1 | q_1 \ldots q_t$ we know $p_1 | q_1$ or $p_1 | q_2 \ldots q_t$ by (82). If $p_1 | q_1$, then $p_1 = q_1$ and the claim is proved. If $p_1 | q_2 \ldots q_t$, then $p_1 | q_2$ or $p_1 | q_3 \ldots q_t$, again by (82). We see there must exist $k \leqslant t$ such that $p_1 | q_k$, and so $p_1 = q_k$. Cancellation yields $p_2 \ldots p_s = q_1 \ldots q_{k-1} q_{k+1} \ldots q_t$. In the same way p_2 must equal one of the remaining q's. Cancel this pair. We continue in this way. Clearly both sides give out at the same time. Thus we have matched up each p with an equal q, and vice versa.

(85) *Note.* We have shown above not only that the same primes must occur on both sides of the equation, but also that a prime that is a repeated factor on one side occurs exactly the same number of times on the other. This theorem and (72) are usually combined as

(86) **The Fundamental Theorem of Arithmetic.** Any integer greater than 1 may be factored into primes. This factorization is unique up to order.

(87) **Exercise.** Suppose $p_1^{q_1} \ldots p_s^{a_s} = q_1^{b_1}, \ldots, q_t^{b_t}$, where the p's and q's are prime, the a's and b's are positive integers, and $p_1 < p_2 < \cdots < p_s$ and $q_1 < q_2 < \cdots < q_t$. Show that $s = t$ and that $p_i = q_i$ and $a_i = b_i$ for $i = 1, 2, \ldots, s$.

(88) **Exercise.** Suppose a and d are positive integers and $d | a$. Exactly what integers b can be expressed in the form $b = (d - xa)/y$ for some integers x and y?

(89) **Exercise*.** Suppose a and b are positive integers. Let the distinct primes dividing a or b or both be $p_1, p_2, \ldots, p_t$. For each $i = 1, 2, \ldots, t$ let α_i and β_i be the number of times p_i appears in the factorization of a and b, respectively; and let M_i be the maximum of α_i and β_i and m_i the minimum. (Some of the α's and β's may be 0.) Prove that $[a,b] = p_1^{M_1} \ldots p_t^{M_t}$ and $(a,b) = p_1^{m_1} \ldots p_t^{m_t}$.

(90) **Exercise.** Use the preceding exercise to prove (27).

(91) **Definition.** Suppose p is a prime. We say p^k *divides a exactly*, and write $p^k \| a$, in case $p^k | a$ while $p^{k+1} \nmid a$.

(92) **Examples.** $3 \| 6$, $2^5 \| 96$, $3^2 \| {-}18$, $3^2 \nmid 27$, $2^2 \nmid 24$, $2^2 \nmid 62$. In the notation of (89), $p_i^{\alpha_i} \| a$ for $i = 1, 2, \ldots, t$.

(93) **True-False.** Here p is a prime, a and b are nonzero integers, and α, β, and k are positive integers, $\alpha < \beta$.

(a) If $p^\alpha \| a$ and $p^\beta \| b$, then $p^{\alpha+\beta} \| ab$.
(b) If $p^\alpha \| (a,b)$, then $p^{2\alpha} \| ab$.
(c) If $p^\alpha \| a$ and $(a,b) = 1$, then $p^\alpha \| ab$.
(d) If $p^\alpha \| a$ and $(a,b) > 1$, then $p^\alpha \nmid ab$.
(e) If $p^\beta \| a$ and $p^\alpha \| (a,b)$, then $p^{\alpha+\beta} \| ab$.
(f) If $p^\alpha \| a$ and $p^\beta \| (a,b)$, then $a + b$ is prime.
(g) If $p^\alpha \| a$, then $p^{k\alpha} \| a^k$.
(h) $(a^k, b^k) = (a,b)^k$.
(i) If $p^\alpha \| a$ and $p^\alpha \| b$, then $p^\alpha \| a + b$.
(j) If $p^\alpha \| a$ and $p^\beta \| b$, then $p^\alpha \| a + b$.
(k) If $p^\alpha \| a$ and $p^\alpha \| b$, then there exists an integer γ such that $\gamma \geqslant \alpha$ and $p^\gamma \| a + b$.

(94) **Exercise*.** Show that $a|b$ if and only if whenever p is a prime and $p^n | a$, then $p^n | b$.

9 PROPERTIES OF DIGITS

(95) Exercises (89) and (94) show how the concepts of divisibility, greatest common divisor, and least common multiple may be viewed from the vantage point of prime factorization. This has practical as well as theoretical advantages. For example, to compute $(450,1000)$ we note that $450 = 2^1 \cdot 3^2 \cdot 5^2$ and $1000 = 2^3 \cdot 5^3 \cdot = 2^3 \cdot 3^0 \cdot 5^3$. Thus [following (89)], $(450,1000) = 2^1 \cdot 3^0 \cdot 5^2 = 50$. In this instance the calculation is not much shorter than using the Euclidean Algorithm. Noting that $512 = 2^9$ while 1001 is odd, on the other hand, enables us to write immediately $(512,1001) = 1$, saving a five-step calculation by the algorithm.

(96) The first step in any application such as the above is the decomposition into primes of the numbers involved. The usual method is to test for divisibility by each prime in order, starting with the smallest. Ordinary long division of n by p tells whether $p|n$; the answer is yes if and only if there is no remainder. Shortcuts exist in the case of a few primes. In the preceding paragraph, for example, we stated that 1001 was odd. The reader no doubt assented to this without any question, relying on what might almost be called a "theorem of the unconscious," namely that a number is even if and only if its last digit is. In spite of the fact that we are already convinced of the truth of this proposition, let us attempt to write out a formal proof, in the hope that it may suggest similar facts about other divisors.

Suppose the positive integer n has the digits $a_t, a_{t-1}, \ldots, a_0$, starting from the left, so $n = a_t 10^t + \cdots + a_1 10^1 + a_0$. For example $564 = 5 \cdot 10^2 + 6 \cdot 10 + 4$, so $t = 2$, $a_2 = 5$, $a_1 = 6$, and $a_0 = 4$. We want to show that n and a_0

are either both even or both odd. Another way to say this is that $n \equiv a_0$ (mod 2). But $n - a_0 = a_t 10^t + \cdots + a_1 10$, which is clearly a multiple of 2. Thus the proposition is proved.

(97) The reader may have noticed that the proof given above depends on the fact that 10 is the base of our number system. This is involved in all the special divisibility tests, and if mankind used a different base than 10 a different set of theorems would apply. Another example is the widely known fact that an integer is divisible by 3 if and only if the sum of its digits has this property. By analogy with the proof just developed we might hope to show that any integer is congruent modulo 3 to the sum of its digits, for then one of these numbers would have remainder 0 when divided by 3 if and only if the other did. (Of course this is more than is really needed to prove what we want. For example x is divisible by 3 if and only if $-x$ is, yet $x \equiv -x$ (mod 3) is not always true.) Let $n = a_t 10^t + \cdots + a_0$ as before. Then

$$n - (a_t + \cdots + a_0) = (10^t - 1)a_t + \cdots + (10 - 1)a_1.$$

It suffices to show this is divisible by 3. But

$$10 - 1 = 9 = 3 \cdot 3, \quad 10^2 - 1 = 99 = 3 \cdot 33,$$

and in general

$$10^k - 1 = 9 \cdot 10^{k-1} + \cdots + 9 = 3(3 \cdot 10^{k-1} + \cdots + 3).$$

Thus our result is proved.

(98) **Exercise***. Show that n is divisible by 5 if and only if its last digit is.

(99) **Exercise***. Show that n is divisible by 9 if and only if the sum of its digits is. (Even though 9 is not prime, this result is useful as it enables us to factor out 3's two at a time.)

(100) **Exercise**. Give a test for the divisibility of n by 3 depending on the digits of n when written to base 9 (i.e., $n = a_t 9^t + \cdots + a_0$, where $0 \leqslant a_i < 9$ for $i = 0, 1, \ldots, t$). Give a test for divisibility by 2, depending on the digits of n written to base 9.

(101) **Exercise**. Prove the polynomial identity $x^k - 1 = (x - 1)(x^{k-1} + x^{k-2} + \cdots + 1)$. Use this to show $9 | 10^k - 1$ for all positive integers k.

(102) **Exercise***. Let m be a positive integer. Show
(a) If $a \equiv a'$ (mod m) and $a' \equiv a''$ (mod m), then $a \equiv a''$ (mod m).
(b) If $a \equiv a'$ (mod m) and $b \equiv b'$ (mod m), then $a + b \equiv a' + b'$ (mod m).
(c) If $a \equiv a'$ (mod m) and $b \equiv b'$ (mod m), then $ab \equiv a'b'$ (mod m).

(d) If $a \equiv b \pmod{m}$, then $a^k \equiv b^k \pmod{m}$ for all positive integers k.

(e) If $a \equiv b \pmod{m}$ and $d \mid m, d > 0$, then $a \equiv b \pmod{d}$.

(103) **Exercise.** Use (102) to show that $9 \mid 10^k - 1$ for all positive integers k. [*Hint*: $10 \equiv 1 \pmod 9$.]

(104) **Exercise***. Show that $n = a_t 10^t + \cdots + a_0$ is divisible by 11 if and only if $a_0 - a_1 + a_2 - a_3 + \cdots$ is. [*Hint*: $10 \equiv -1 \pmod{11}$.]

(105) We illustrate the above by factoring $n = 37719$. Since $3 + 7 + 7 + 1 + 9 = 27$. We see $9 \mid n$ by (99). Division produces $37719 = 9 \cdot 4191$. Now we concentrate on 4191. Since $4 + 1 + 9 + 1 = 15$, this is divisible by 3 but not by 9 by (97) and (99). We have $4191 = 3 \cdot 1397$. Ordinary division shows $7 \nmid 1397$. The next prime is 11. By (104) this is a factor since it divides $7 - 9 + 3 - 1 = 0$. In fact $1397 = 11 \cdot 127$. Since $11 \nmid 7 - 2 + 1$ we go on to 13. But $13^2 = 169$, so $127/13 < 13$. Thus if 127 had a proper factor $\geqslant 13$ it would also have a factor < 13. Since the latter possibility has been eliminated, we conclude that 127 is a prime. Thus $37719 = 3^3 \cdot 11 \cdot 127$.

(106) The argument given at the end of the previous paragraph may be generalized as follows. The proof is left to the reader

(107) **THEOREM.** If the positive integer $n > 1$ has no prime factors $\leqslant \sqrt{n}$, then it is prime.

(108) **Exercise.** Write each of the following numbers in the form $p_1^{a_1} \cdots p_t^{a_t}$, where the p's are prime, the a's are positive integers, and $p_1 < p_2 < \cdots < p_t$: 293, 1001, 1763, 2310, 6561, 1000000.

10 THE SIEVE OF ERATOSTHENES

(109) Suppose we want to determine all the primes less than 100. We might proceed as follows. First we write out the integers from 2 to 100. We know 2 is prime; let us circle it and cross out the remaining even numbers on our list. The next lowest number which hasn't been crossed out is 3; we circle it and cross out every third number thereafter, since 3 is a proper divisor of each of these. Our list now looks as follows:

② ③ 4̸ 5 6̸ 7 8̸ 9̸ 1̸0̸ 11 1̸2̸ 13 1̸4̸ 1̸5̸ 1̸6̸ 17 $\cdots$

At each stage in this procedure the smallest number which has not been circled or crossed out must be prime, since otherwise it would have a smaller prime divisor, and so have been eliminated already. This process is called the "sieve of Eratosthenes."

(110) **Exercise.** Use the technique of (109) to find the 4 primes between 1000 and 1030.

(111) The operation of the sieve of Eratosthenes suggests that primes should be rarer among the larger integers. For example, its application to the numbers between 100 and 150 consists in crossing out the multiples of the 25 primes less than 100. Applying the sieve between 1000 and 1050, however, we eliminate not only the multiples of all these primes, but also the multiples of the 143 additional primes between 100 and 1000. It turns out that there are 10 primes between 100 and 150, but only 8 between 1000 and 1050. Between 10,000 and 10,050 there are just 4 primes.

   These considerations suggest the possibility that at some stage in the application of the sieve all larger numbers will have been crossed out, so that there would be no more primes. This would mean that there would exist only finitely many primes, say $p_1, p_2, \ldots, p_t$, and each integer greater than 1 could be written as a product of powers of these primes. The integer $P = p_1 p_2 \ldots p_t$ would then have the interesting property that $(P,n) > 1$ whenever $n > 1$, since all the distinct prime factors of n would appear in P. Let us make a table of (P,n) for small values of n.

n	(P,n)
1	1
2	2
3	3
4	2
5	5
6	6
7	7
8	2
9	3
10	10

Back in (61) we made a similar table for $(6,n)$. It turned out to be periodic, with period 6. In fact we proved the theorem that $b \equiv b'$ (mod a) implies $(a,b) = (a,b')$. But this doesn't square with what we know about (P,n). The theorem says that if $n \equiv 1$ (mod P), then $(P,n) = (P,1) = 1$; for example $(P,P+1) = 1$. Our definition of P, however, led us to the conclusion that $(P,n) > 1$ for $n > 1$. The assumption that the number of primes is finite has produced a contradiction.

(112) **THEOREM.** There exist infinitely many primes.

(113) **Exercise.** Pretend that the number P defined in (111) exists and extend the table given there to $n = 11, 12, \ldots, 20$. You're living a lie.

11 A FORMULA FOR τ

(114) We now return to the evaluation of $\tau(n)$, sidetracked when we got into the much more interesting problem of prime factorization. What we know now makes the problem easy. Suppose $n = p_1^{\alpha_1} \ldots p_t^{\alpha_t}$, where the p's are distinct primes. If d is a positive divisor of n, then $n = dd'$, and so by unique factorization $d = p_1^{\beta_1} \ldots p_t^{\beta_t}$, where $0 \leqslant \beta_i \leqslant a_i$, for $i = 1, 2, \ldots, t$. (We allow the β's to equal 0 to take care of primes not dividing d at all.) For example, if $n = 63 = 3^2 \cdot 7$, then the positive divisors of n are

$$3^0 7^0 \qquad 3^1 7^0 \qquad 3^2 7^0$$
$$3^0 7^1 \qquad 3^1 7^1 \qquad 3^2 7^1.$$

Returning to the general case, we note that unique factorization also tells us that each different choice for β_1 through β_t gives us a different d. Since there are $a_1 + 1$ possibilities for β_1, namely $0, 1, \ldots,$ and $a_1, a_2 + 1$ possibilities for β_2, etc., we have

(115) **THEOREM.** If $n = p_1^{\alpha_1} \ldots p_t^{\alpha_t}$, where the p's are distinct primes, then $\tau(n) = (\alpha_1 + 1)(\alpha_2 + 1) \ldots (\alpha_t + 1)$.

(116) **Exercise.** Write out the positive divisors of 360. Evaluate $\tau(n)$ for
$$n = 210, 243, 244, \text{ and } 1,000,000.$$

(117) When, in (114), we wrote out the positive divisors of 63, we found it convenient to organize them into a rectangular array. This array may have reminded the reader of a multiplication table; indeed, that is exactly what it was. Let us look at it again.

	1	3	9
1	1	3	9
7	7	21	63

We have written the positive divisors of 9 along the top and those of 7 along the left side. Each entry in the table is the product of a divisor of 9 and one of 7. Thus in this case $\tau(63) = \tau(9)\tau(7) = 3 \cdot 2 = 6$. This suggests that we might prove in general that $\tau(ab) = \tau(a)\tau(b)$ by showing that the divisors of ab are just all products of a divisor of a with one of b.

Let us suppose a and b have positive divisors $d_1, d_2, \cdots, d_s$ and $e_1, e_2, \cdots, e_t$ respectively. Thus $\tau(a) = s$ and $\tau(b) = t$. We consider the array

$$d_1 e_1 \qquad d_2 e_1 \quad \cdots \quad d_s e_1$$
$$d_1 e_2 \qquad d_2 e_2 \quad \cdots \quad d_s e_2$$
$$\cdots \cdots \cdots \cdots$$
$$d_1 e_t \qquad d_2 e_t \quad \cdots \quad d_s e_t.$$

This array has st entries, so in order to show $\tau(ab) = st$ we must demonstrate three things: (1) every entry is a divisor of ab, (2) every positive divisor of ab appears in the array, and (3) no two entries are equal. Since (1) is easy to see we proceed to (2). Suppose c is a positive divisor of ab. Then there exists an integer k such that $ab = kc$. Imagine the prime factorization of both sides of this equation; by the Fundamental Theorem of Arithmetic we can match up the primes on both sides in a one-to-one fashion. In particular, some of the primes in the factorization of c match up with primes in the factorization of a, the rest with primes in b. We see that c is a product of a divisor of a with one of b.

We now turn to condition (3). Suppose that two entires of the array are equal, say $d_i e_j = d_k e_l$. If we knew that the prime factors of d_i could only occur in d_k, and vice versa, we could conclude that $d_i = d_k$, and $e_j = e_l$ would follow. Unfortunately this need not be true. For example, if p is a prime dividing both a and b, then $p \cdot 1$ and $1 \cdot p$ will appear at different places in the array. Evidently we need an additional hypothesis in order to prove (3), namely that no prime divides both a and b. If this is the case we easily see that $d_i e_j = d_k e_l$ implies that $d_i = d_k$ and $e_j = e_l$, since, for example, the primes dividing d_i cannot appear in e_l, a divisor of b, and so must all turn up in d_k. Since saying no prime divides both a and b is equivalent to saying $(a,b) = 1$, we have proved

(118) **THEOREM**. If a and b are relatively prime, then $\tau(ab) = \tau(a)\tau(b)$.

(119) This theorem could have been proved much more easily directly from the formula we developed for $\tau(n)$. We have proved somewhat more than (118), however, namely

(120) **THEOREM**. If $(a,b) = 1$, the set of positive divisors of ab consists exactly of all products de, where d is a positive divisor of a and e is a positive divisor of b. Furthermore, these products are all distinct.

(121) **Exercise**. Prove (118) directly from (115).

2

Numerical Functions

Functions like τ, defined on the positive integers, are variously called numerical, arithmetic, or number-theoretic functions. They are treated in this chapter.

12 MULTIPLICATIVE FUNCTIONS

(122) **Definition.** Suppose f is a function defined on the positive integers. We say that f is *multiplicative* in case whenever a and b are relatively prime positive integers, then $f(ab) = f(a)f(b)$.

(123) Many of the functions that arise naturally in number theory turn out to be multiplicative. The advantage in knowing that a function is multiplicative is that then the problem of its determination is reduced to that of evaluating it at prime powers. This is because if f is multiplicative and $n = p_1^{\alpha_1} \cdots p_t^{\alpha_t}$, where the p's are distinct primes, then $f(n) = f(p_1^{\alpha_1} \cdots p_t^{\alpha_t}) = f(p_1^{\alpha_1})f(p_2^{\alpha_2} \cdots p_t^{\alpha_t}) = \cdots = f(p_1^{\alpha_1})f(p_2^{\alpha_2}) \cdots f(p_t^{\alpha_t})$. For example, it is easy to see that the positive divisors of p^α are exactly $1, p, p^2, \ldots, p^\alpha$. Thus $\tau(p^\alpha) = \alpha + 1$. The formula of (115) then follows immediately from the fact that τ is multiplicative.

As another example let us consider the function σ, where $\sigma(n)$ is defined to be the *sum* of the positive divisors of n. For instance $\sigma(7) = 1 + 7 = 8$, $\sigma(9) = 1 + 3 + 9 = 13$, and $\sigma(63) = 1 + 3 + 7 + 9 + 21 + 63 = 104 = \sigma(7)\sigma(9)$. This and similar examples lead us to hope that σ is multiplicative; let us attempt a proof. Assume $(a,b) = 1$. Let a and b have the positive divisors $d_1, d_2, \ldots, d_s$ and $e_1, e_2, \ldots, e_t$ respectively. Then $\sigma(a) = d_1 + \cdots + d_s$ and $\sigma(b) = e_1 + \cdots + e_t$. We must show that $\sigma(ab) = (d_1 + \cdots + d_s)(e_1 + \cdots + e_t)$. But multiplying out this last expression gives us

$$d_1e_1 + d_2e_1 + \cdots + d_se_1$$
$$+ d_1e_2 + \cdots + d_se_2$$
$$+ \cdots +$$
$$+ d_1e_t + \cdots + d_se_t,$$

which, according to (120), is exactly the sum of the positive divisors of ab. Thus we have proved that σ is multiplicative.

To find a formula for σ it only remains to evaluate $\sigma(p^\alpha)$, where p is a prime. This is just $1 + p + p^2 + \cdots + p^\alpha$, a geometric progression. The standard formula (see (125)) gives $\sigma(p^\alpha) = (p^{\alpha+1} - 1)/(p - 1)$.

(124) **THEOREM.** If $n = p_1^{\alpha_1} \cdots p_t^{\alpha_t}$, where the p's are distinct primes, then

$$\sigma(n) = \frac{p_1^{\alpha_1+1} - 1}{p_1 - 1} \cdots \frac{p_t^{\alpha_t+1} - 1}{p_t - 1}.$$

(125) **Exercise*.** Show that if a and r are any real numbers, $r \neq 1$, then
$a + ar + ar^2 + \cdots + ar^n = a(r^{n+1} - 1)/(r - 1)$. [*Hint*: Call the sum S. Show $rS - S = ar^{n+1} - a$.]

(126) **Exercise.** Evaluate $\sigma(40)$ both directly from the definition and by use of (124). Evaluate $\sigma(10^6)$ and $\sigma(360)$.

(127) **Definition.** We define $\sigma_k(n)$ to be the sum of the kth powers of the positive divisors of n. In particular, $\sigma_1(n) = \sigma(n)$ and $\sigma_0(n) = \tau(n)$ as defined previously.

(128) **Examples.** $\sigma_2(63) = 1^2 + 3^2 + 7^2 + 9^2 + 21^2 + 63^2 = 4550$, $\sigma_3(6) = 1^3 + 2^3 + 3^3 + 6^3 = 252$, $\sigma_0(20) = 1^0 + 2^0 + 4^0 + 5^0 + 10^0 + 20^0 = 6$.

(129) **Exercise.** Evaluate $\sigma_2(7)$, $\sigma_2(9)$, $\sigma_0(900)$, $\sigma_4(4)$.

(130) **Exercise.** Prove that σ_2 is a multiplicative function. [*Hint*: Copy the argument used on σ in (123).]

(131) **Exercise.** Show that $\sigma_2(p^\alpha) = (p^{2\alpha+2} - 1)/(p^2 - 1)$ if p is prime.

(132) **Exercise.** Prove a formula for $\sigma_2(n)$.

(133) **Exercise.** Prove that if the p's are distinct primes, then $\sigma_k(p_1^{\alpha_1} \cdots p_t^{\alpha_t}) = \dfrac{p_1^{k(\alpha_1+1)} - 1}{p_1^k - 1} \cdots \dfrac{p_t^{k(\alpha_t+1)} - 1}{p_t^k - 1}.$

(134) **Exercise.** In the definition of σ_k we need not restrict k to nonnegative integers; it may be any real number. Show that if n is any positive integer, then $n\sigma_{-1}(n) = \sigma(n)$. [*Hint*: Show that as d runs through the positive divisors of n, so does n/d.]

13 THE SUMMATION AND PRODUCT NOTATIONS

(135) The reader is probably already familiar with the Σ notation for sums, by which $f(1) + f(2) + \cdots + f(n)$ is expressed in the more compact form $\Sigma_{i=1}^n f(i)$. For example $\Sigma_{i=1}^4 i^2 = 1^2 + 2^2 + 3^2 + 4^2 = 30$. In number theory this conven-

tion is often extended by simply writing under the Σ the condition or conditions delimiting the numbers over which the summation is to be made. For example $\Sigma_{d|5}\, d^2$ denotes the sum of the squares of the divisors of 5, so $\Sigma_{d|6}\, d^2 = 1^2 + (-1)^2 + 5^2 + (-5)^2 = 52$. Likewise, $\Sigma_{d|5 \atop d>0}\, d = 1 + 2 + 3 + 6 = 12$, and, in general, $\Sigma_{d|n \atop d>0}\, d = \sigma(n)$. We could have defined $\sigma_k(n)$ as $\Sigma_{d|n \atop d>0}\, d^k$. The "dummy variable" is ordinarily understood to be restricted to integral values; thus by $\Sigma_{n^2<5}\, |n|$ we mean $|-2| + |-1| + |0| + |1| + |2|$. When $S(n)$ is some statement involving n the expression $\Sigma_{S(n)}\, 1$ merely counts the integers for which $S(n)$ is true. For example, $\Sigma_{0<n\leqslant5}\, 1 = 4$, and $\Sigma_{d|n \atop d>0}\, 1 = \tau(n)$. It is standard to make the convention that $\Sigma_{S(n)} f(n) = 0$ in case $S(n)$ is true for no integers.

The user of this notation should always make clear which variable the summation runs over. We illustrate the importance of this by listing four true statements:

(a) If x is an integer, $0 \leqslant x \leqslant 3$, then $\Sigma_{|x|+|y|\leqslant3}\, 1 = 7 - 2x$.
(b) If y is an integer, $0 \leqslant y \leqslant 3$, then $\Sigma_{|x|+|y|\leqslant3}\, 1 = 7 - 2y$.
(c) If $x = -1$ and $y = 7$, then $\Sigma_{|x|+|y|<3}\, 1 = 0$.
(d) $\Sigma_{|x|+|y|\leqslant3}\, 1 = 16$.

In (a) y is the dummy variable; in (b) it is x. In (c) there is no variable of summation. In (d), finally, both x and y are variables, and the sum counts the *pairs* x,y such that $|x| + |y| \leqslant 3$. A good general rule to remember is that a dummy variable has no meaning outside its summation. Thus $\Sigma_{d|64} \tau(d) = (d^2 - 1)/d$ is nonsense on the face of it, since the d has strayed from the Σ.

Corresponding to the Σ notation for sums is the Π notation for products. Thus $\Pi_{i=1}^n f(i) = f(1)f(2) \ldots f(n)$, $\Pi_{i=0}^2 (2i + 1) = 1 \cdot 3 \cdot 5$, and $\Pi_{p|12 \atop p\ prime}\, p = 2 \cdot 3$.

In this notation the conclusion of (124) could be expressed as

$$\sigma\left(\prod_{i=1}^t p_i^{\alpha_i}\right) = \prod_{i=1}^t \frac{p_i^{\alpha_i+1} - 1}{p_i - 1}.$$

The convention is made that $\Pi_{S(n)} f(n) = 1$ in case $S(n)$ is never true.

(136) **Exercise.** Evaluate $\Sigma_{d|6}\, d$, $\Sigma_{d|15 \atop d<0} (15/d)$, $\Sigma_{p^2||20 \atop p\ prime} p$, $\Sigma_{p|36 \atop p\ prime} 1$, $\Sigma_{p||36 \atop p\ prime} 1$, $\Sigma_{0<xy<5}\, x$, $\Pi_{i=1}^5 i^2$, $\Pi_{d|6 \atop d>0}\, d$, $\Pi_{|d|<10^6}\, d$, $\Pi_{5|d \atop 0<d<5}\, d^2$.

(137) **Exercise*.** Suppose $S(n)$ is a statement involving n true for exactly s values of n. Show

$$\Sigma_{S(n)} af(n) = a\, \Sigma_{S(n)} f(n);$$
$$\Sigma_{S(n)} f(n) + \Sigma_{S(n)} g(n) = \Sigma_{S(n)}(f(n) + g(n));$$
$$\Sigma_{S(n)} a = sa;$$
$$\Pi_{S(n)} af(n) = a^s\, \Pi_{S(n)} f(n);$$

$$(\Pi_{S(n)} f(n))\,(\Pi_{S(n)} g(n)) = \Pi_{S(n)} f(n) g(n);$$
$$\Pi_{S(n)}(f(n))^k = (\Pi_{S(n)} f(n))^k.$$

(138) **Exercise***. Suppose $S(n)$ and $T(n)$ are statements involving n, each true for only finitely many values of n. Let f be a function of two variables. Show that $\Sigma_{S(x)}(\Sigma_{T(y)} f(x,y)) = \Sigma_{T(y)}(\Sigma_{S(x)} f(x,y)) = \Sigma_{S(x)\,\&\,T(y)} f(x,y)$. Show that if $f(x,y) = g(x)h(y)$, then all these sums are also equal to $(\Sigma_{S(x)} g(x)) \cdot (\Sigma_{T(y)} h(y))$.

14 A THEOREM ABOUT MULTIPLICATIVE FUNCTIONS

(139) Exercises (130) and (133) may lead the reader to wonder how far the argument at the end of (123) can be pushed. Not only is the sum of the positive divisors of n a multiplicative function of n; so is the sum of the squares, or, in fact, kth powers of these divisors. In general, what can we do to the positive divisors of n so as to have the sum of the results a multiplicative function of n? Posed less awkwardly, the question is for which functions f the function F, defined by $F(n) = \Sigma_{\substack{d\mid n \\ d>0}} f(d)$, is a multiplicative function of n. Suppose a and b are relatively prime, with positive divisors $d_1,\ldots,d_s$ and $e_1,\ldots,e_t$ respectively. We want $F(ab) = F(a)F(b)$. The latter quantity is

$$f(d_1) + \cdots + f(d_s)\,(f(e_1) + \cdots + f(e_t)) = \sum_{\substack{1 \leqslant i \leqslant s \\ 1 \leqslant j \leqslant t}} f(d_i)f(e_j).$$

(See 138.) On the other hand (120) tells us that $F(ab) = \Sigma_{\substack{1 \leqslant i \leqslant s \\ 1 \leqslant j \leqslant t}} f(d_i e_j)$. Thus to prove that F is multiplicative it suffices to show that $f(d_i)\,f(e_j) = f(d_i e_j)$ for all i and j. Since no prime divides both a and b, the same can be said of any d and e; thus $(d_i, e_j) = 1$ for all i and j. We see that what is needed is just that f itself be multiplicative. We have proved

(140) **THEOREM.** If f is a multiplicative function and $F(n) = \Sigma_{\substack{d\mid n \\ d>0}} f(n)$, then F is also multiplicative.

(141) **Examples.** Let $f(n) = n$ for all n. Then $f(ab) = ab = f(a)f(b)$ for all a and b. Thus $\Sigma_{\substack{d\mid n \\ d>0}} d = \sigma(n)$ is a multiplicative function.

Let $f(n) = 1$ for all n. This is clearly multiplicative. Thus $\Sigma_{\substack{d\mid n \\ d>0}} 1 = \tau(n)$ is multiplicative.

Let $f(n) = n^k$. Then $f(ab) = (ab)^k = a^k b^k = f(a)f(b)$. Thus $\Sigma_{\substack{d\mid n \\ d>0}} d^k = \sigma_k(n)$ is multiplicative.

Let $f(n) = \tau(n)$. Then the function $F(n) = \Sigma_{\substack{d\mid n \\ d>0}} \tau(d)$ is multiplicative, since τ is.

(142) **Exercise.** Let F be defined as at the end of section (141). Show that if the p's are distinct primes, then $F(p_1^{\alpha_1} \ldots p_t^{\alpha_t}) = 2^{-t} \Pi_{1 \leqslant i \leqslant t} (\alpha_i + 1)(\alpha_i + 2)$. [*Hint:* Recall $1 + 2 \cdots + n = n(n+1)/2$.]

15 PERFECT NUMBERS

(143) Perhaps the reader is wondering "what good" a function like σ is. Who cares that we have found a formula for the sum of the positive divisors of n? This is a question that each person must answer for himself, since number theory, like mountain climbing, is an endeavor practiced entirely for its own sake. Among the reasons why some people do care about it might be:

1. *Accomplishment.* Solving a mathematical problem may give the same pleasant feeling of success as finishing a difficult crossword puzzle or winning a game of tennis.

2. *Understanding.* Knowing *why* a number is divisible by 9 if and only if the sum of its digits is (and thus the justification of the old check of "casting out 9's") may bring satisfaction.

3. *Skill.* A person might take pride in being able to play the guitar or water ski, or, just as well, count the divisors of a million in a few seconds or compute the greatest common divisor of two large numbers.

4. *Beauty.* Many find beauty in number theory, either in the structure of the integers themselves (as an astronomer might find beauty in the laws of the universe), or in an ingenious or elegant proof.

The study of numbers for their own sake is by no means new. The Greeks were interested in numbers like 6, which is the sum of its positive divisors other than itself; $6 = 1 + 2 + 3$. If n is any number with this property, then $n = \sigma(n) - n$, since by definition σ adds in the divisor n itself.

(144) **Definition.** We say n is a *perfect* number in case $\sigma(n) = 2n$.

(145)

n	$\sigma(n)$	n	$\sigma(n)$	n	$\sigma(n)$
1	1	11	12	21	32
2	3	12	28	22	36
3	4	13	14	23	24
4	7	14	21	24	60
5	6	15	24	25	31
6	12	16	31	26	42
7	8	17	18	27	40
8	15	18	39	28	56
9	13	19	20	29	30
10	18	20	42	30	72

Construction of the above table was simplified by using a few properties of σ. The primes were easily filled in, since $\sigma(p) = p + 1$. Prime powers follow the rule that $\sigma(p^\alpha) = \sigma(p^{\alpha-1}) + p^\alpha$. Thus $\sigma(4) = \sigma(2) + 4$, $\sigma(8) = \sigma(4) + 8$, etc. The remaining values were then determined by the fact that σ is multiplicative. For example $\sigma(6) = \sigma(2)\sigma(3) = 12$, and $\sigma(18) = \sigma(2)\sigma(9) = 39$. (But $\sigma(3)\sigma(6) = 48$. What's wrong?)

(146) **Exercise.** Extend the above table to $n = 40$.

(147) Our table has yielded only one more perfect number, 28. Although two numbers are certainly not much to go on, perhaps by trying to understand *why* 6 and 28 are perfect we can construct others. The formula of (124) is a natural starting point. We calculate

$$\sigma(6) = \sigma(2 \cdot 3) = \left(\frac{2^2 - 1}{2 - 1}\right)\left(\frac{3^2 - 1}{3 - 1}\right) = 3 \cdot 2 \cdot 2.$$

Likewise,

$$\sigma(28) = \sigma(2^2 \cdot 7) = \left(\frac{2^3 - 1}{2 - 1}\right)\left(\frac{7^2 - 1}{7 - 1}\right) = 7 \cdot 2 \cdot 4.$$

Let us write out these equations again, using arrows to show how the factors of n contribute to $\sigma(n)$.

$$\sigma(2 \cdot 3) = 3 \cdot 2 \cdot 2$$
$$\sigma(2^2 \cdot 7) = 7 \cdot 2 \cdot 2^2.$$

We see that both 6 and 28 are of the form $2^\alpha b$, b odd, where $\sigma(2^\alpha) = b$ and $\sigma(b) = 2 \cdot 2^\alpha$. Are there any other perfect numbers of this form? By (124) $b = \sigma(2^\alpha) = (2^{\alpha+1} - 1)/(2 - 1) = 2^{\alpha+1} - 1 = \sigma(b) - 1$. Thus $\sigma(b) = b + 1$. But this happens if and only if the positive divisors of b are exactly b and 1, i.e., if and only if b is prime.

(148) **THEOREM.** Suppose $n = 2^\alpha(2^{\alpha+1} - 1)$, where $2^{\alpha+1} - 1$ is prime. Then n is perfect.

Proof. (Just to make sure we have everything straight.) Clearly $(2^\alpha, 2^{\alpha+1} - 1) = 1$. Thus by (124) $\sigma(n) = \sigma(2^\alpha)\sigma(2^{\alpha+1} - 1) = (2^{\alpha+1} - 1)(2^{\alpha+1} - 1 + 1)$, since $2^{\alpha+1} - 1$ is prime. We see $\sigma(n) = (2^{\alpha+1} - 1) \cdot 2 \cdot 2^\alpha = 2\sigma(n)$.

(149) The question becomes when $2^{\alpha+1} - 1$ is prime.

α	$2^{\alpha+1} - 1$
1	3
2	7
3	15
4	31

Since 31 is prime $2^4 \cdot 31 = 496$ is another perfect number.

(150) **Exercise.** Find the next perfect number of the form $2^\alpha(2^{\alpha+1} - 1)$.

(151) We have proved that $2^\alpha(2^{\alpha+1} - 1)$ is perfect whenever $2^{\alpha+1} - 1$ is prime. A natural question at this point is whether all perfect numbers have this form. This wasn't proved in (147), since there we started with certain assumptions about how $\sigma(n)$ was put together. It seems reasonable to try a proof along the same lines, however. Let us assume that n is perfect. Since the prime 2 seems to play a special role, let $n = 2^\alpha b$, where b is odd and we allow the possibility that $\alpha = 0$. Then $\sigma(n) = \sigma(2^\alpha)\sigma(b) = (2^{\alpha+1} - 1)\sigma(b)$, by (124). Since n is perfect, $(2^{\alpha+1} - 1)\sigma(b) = 2 \cdot 2^\alpha b = 2^{\alpha+1} b$. We would like to show b to be prime. We *do* notice that the ratio of b to $\sigma(b)$ appears near to 1, since $b/\sigma(b) = (2^{\alpha+1} - 1)/2^{\alpha+1}$. This is promising, since $\sigma(b)$ is closest to b when b is prime. If we could conclude that $b = 2^{\alpha+1} - 1$ and $\sigma(b) = 2^{\alpha+1}$, we would be home free, but fractions may be equal without their numerators and denominators being equal. Although $(2^{\alpha+1} - 1)/2^{\alpha+1}$ is in lowest terms, $b/\sigma(b)$ may not be. Let $(b,\sigma(b)) = c$. Then $(b/c)/(\sigma(b)/c)$ *is* in lowest terms, so $b/c = 2^{\alpha+1} - 1$ and $\sigma(b)/c = 2^{\alpha+1}$. We see $b = c(2^{\alpha+1} - 1) = \sigma(b) - c$, or $\sigma(b) = b + c$. But c divides b, thus the last equation implies that b and c are the *only* positive divisors of b; i.e., $c = 1$ and b is prime. Since $c = 1$, we have $b = 2^{\alpha+1} - 1$. Thus we have proved that each perfect number is of the form $2^\alpha(2^{\alpha+1} - 1)$, where $2^{\alpha+1} - 1$ is prime.

No one who has read this far should fail to read (152).

(152) Numbers of the form $2^\alpha(2^{\alpha+1} - 1)$ with $2^{\alpha+1} - 1$ prime are sometimes called "Euclid numbers," since the proof that such numbers are perfect appears in the *Elements*. Dickson's *History of the Theory of Numbers*, Volume I, chronicles the attention many early writers paid to perfect numbers. Saint Augustine said that God created the world in six days rather than at once because "the perfection of the work is signified by the [perfect] number 6." (It didn't work.) Others asserted that there are infinitely many perfect numbers, that, indeed, there is exactly one between 1 and 10, another between 10 and 100, another between 100 and 1000, etc., that each perfect number is a Euclid number, that they alternately end in the digits 6 and 8, and that $2^{\alpha+1} - 1$ is a prime (and so $2^\alpha(2^{\alpha+1} - 1)$ is perfect) whenever $\alpha + 1$ is odd. None of these assertions has ever been proved.

The last sentence was intended to produce either shock or smug satisfaction in the reader. If he feels neither it may be time to get a cup of coffee and start over at (151). Of course the assertion that each perfect number is a Euclid number is exactly what we purported to prove in (151); and now I say this *hasn't* been proved. This contradiction is easily explained: the proof in (151) was phony. The reader who caught the error there may now congratulate himself again; the reader who was conned should go back and look for it. It will be revealed in (153).

A word is in order about the other unfounded assertions of the early mathematicians (?). The existence of exactly one perfect number between adjacent powers of 10 and the alternation of 6 and 8 as final digits are both false. If $2^k - 1$ were prime for all odd values of k the existence of infinitely many perfect numbers would, of course, follow. Unfortunately $2^9 - 1 = 511 = 7 \cdot 73$. Perhaps the fact that several people got into print with the statement that 511 is prime may make us feel better about our own mistakes. The possibility remains, of course, that $2^k - 1$ might be prime for infinitely many values of k. This has never been proved one way or the other.

(153) The error in the argument of (151) comes when we conclude from $\sigma(b) = b + c$ that $c = 1$ and b is prime. It is true that c is a divisor of b; we had $b = c(2^{\alpha+1} - 1)$. What need not be true is that c is a *different* divisor than b itself. In fact if $\alpha = 0$, then $b = c$, and $\sigma(b) = b + c = 2b$ merely says that b is perfect. If $\alpha > 0$, then $2^{\alpha+1} - 1 > 1$ and so $c \neq b$. Thus we have really proved

(154) **THEOREM.** Every even perfect number is of the form $2^\alpha(2^{\alpha+1} - 1)$, where $2^{\alpha+1} - 1$ is prime.

(155) **Exercise.** Show that if n is an odd perfect number, then $n = pa^2$, where p is prime. [*Hint*: Consider the equation $\sigma(n) = 2n$. When is $\sigma(p^\alpha)$ odd?]

(156) We are left with the possibility that there exist odd perfect numbers not of Euclid's type. Although no one has ever proved odd perfect numbers do not exist, neither has anyone ever found one. It is known that no odd perfect number is less than e^{52729}.

16 MERSENNE NUMBERS

(157) Theorems (148) and (154) together show that the even perfect numbers are in one-to-one correspondence with the primes of the form $2^k - 1$.

k	$2^k - 1$	k	$2^k - 1$
1	1	6	63
2	3	7	127
3	7	8	255
4	15	9	511
5	31	10	1023

The values of k in the above table for which $2^k - 1$ is prime are 2, 3, 5, and 7. That these are precisely the primes less than 10 may be only a coincidence, but we would be foolish not to follow up such an obvious lead. We note that

$$2^4 - 1 = 15 = 3 \cdot 5 = (2^2 - 1)(2^2 + 1).$$

In fact, whenever k is even, say $k = 2t$, then $2^k - 1 = (2^t - 1)(2^t + 1)$, so $2^k - 1$ is composite. The only exception is $k = 2$; then $2^t - 1 = 2^1 - 1 = 1$ so the factorization degenerates. The 2 is really incidental; $2^{2t} - 1 = (2^t - 1)(2^t + 1)$ may be viewed as the special case of the polynomial identity $x^2 - 1 = (x - 1)(x + 1)$ when $x = 2^t$. In the same way we get a factorization of $2^{3t} - 1$ by letting $x = 2^t$ in the identity $x^3 - 1 = (x - 1)(x^2 + x + 1)$. For example $2^9 - 1 = (2^3 - 1)$ $(2^6 + 2^3 + 1) = 7 \cdot 73$. We could prove that in general $k = st$ implied that $2^k - 1$ was composite if we knew that the polynomial $x - 1$ always divided the polynomial $x^s - 1$. Since $x = 1$ is clearly a root of $x^s - 1 = 0$, this follows from the theory of equations, but we can also give a direct proof. From the examples given it appears that

$$x^s - 1 = (x - 1)(x^{s-1} + x^{s-2} + \cdots + 1).$$

But the last factor is a geometric progression; by (125) it equals $(x^s - 1)/(x - 1)$, which proves the identity. Plugging in $x = 2^t$ yields

$$2^{st} - 1 = (2^t - 1)(2^{t(s-1)} + 2^{t(s-2)} + \cdots + 1).$$

In order to conclude that $2^{st} - 1$ is composite we must have that neither of these factors is 1. Is suffices that $t > 1$ and $s > 1$; i.e., that k has a positive factor other than itself and 1. We have proved

(158) **THEOREM**. If $2^k - 1$ is prime, then k is prime.

(159) **Definition.** If p is prime, $2^p - 1$ is called a *Mersenne number*. It is denoted by M_p.

(160) **Exercise.** Prove or disprove that all Mersenne numbers are prime.

(161) **Exercise.** Prove that if $a > 2$ and $k > 1$, then $a^k - 1$ is composite.

(162) **Exercise.** Prove that $3 | 2^k - 1$ if k is even.

(163) **Exercise.** Prove that $3 \nmid 2^k - 1$ if k is odd.

(164) **Exercise.** Does $7 | 2^{49} - 1$? [*Hint*: $2^3 \equiv 1 \pmod 7$.]

17 REVERSING A FORMULA

(165) Since we are so often interested in summing a function over only the *positive* divisors of an integer, we will agree that from now on by $\Sigma_{d|n} f(d)$ we mean $\Sigma_{d|n \atop d>0} f(d)$; i.e., the condition $d > 0$ will be understood.

Back in (140) we proved that if the function f is multiplicative, then so is the function F defined by $F(n) = \Sigma_{d|n} f(d)$. It might very well be that f is easier to prove multiplicative than F, in which case this theorem would be handy. For example the function $f(n) = n$ is clearly multiplicative, so $\sigma(n) = \Sigma_{d|n} d$ inherits this property. It is obvious from the definition of σ what the corresponding f is, but this might not always be so. We are led to the general problem of recovering f, given F. Three questions immediately come to mind:

    1. Given a function F defined on the positive integers, does there necessarily exist a function f such that $F(n) = \Sigma_{d|n} f(d)$ for all n?

    2. If such an f does exist, is it unique?

    3. Is there a formula for f in terms of F?

(166) Let us suppose for the moment that f does exist such that $F(n) = \Sigma_{d|n} f(d)$ for all n. This formula represents infinitely many equations linking f and F. For example $F(1) = \Sigma_{d|1} f(d) = f(1)$. We see that at least $f(1)$ can be determined in terms of F; it must be $F(1)$. In the same way $F(2) = \Sigma_{d|2} f(d) = f(1) + f(2)$. But then $f(2) = F(2) - f(1) = F(2) - F(1)$. Likewise $F(3) = f(1) + f(3)$, so $f(3) = F(3) - F(1)$. Again $F(4) = f(1) + f(2) + f(4) = F(1) + F(2) - F(1) + f(4)$, so $f(4) = F(4) - F(2)$.

(167) **Exercise.** Extend the following table to $n = 10$.

n	$f(n)$
1	$F(1)$
2	$F(2) - F(1)$
3	$F(3) - F(1)$
4	$F(4) - F(2)$.

(168) It appears that the above table can be continued indefinitely. To determine $f(n)$ in terms of F we merely write $F(n) = f(d_1) + f(d_2) + \cdots + f(n)$, where $d_1, d_2, \ldots, n$ are the positive divisors of n. Since for $d < n$ the table already contains $f(d)$ in terms of F, $f(n)$ can also be written in terms of F. Notice that this does not establish the *existence* of f, since this was assumed to begin with. It does say that *if* f exists then it is unique; i.e., 2 is answered. To work the argument given above into a more formal proof we will invoke mathematical induction—in a form with which the reader may or may not be familiar.

(169) **Exercise*.** Prove that the following statements are equivalent:

    (a) Suppose that $S(n)$ is a statement involving n. If $S(1)$ is true, and if $S(n + 1)$ is true whenever $S(n)$ is true, then $S(n)$ is true for all positive integers n.

    (b) Suppose $T(n)$ is a statement involving n. If $T(1)$ is true, and if $T(n + 1)$ is true whenever $T(1), T(2), \ldots, T(n)$ are all true, then $T(n)$ is true for all positive integers n.

    (c) If U is any nonempty set of positive integers, then U contains a smallest element. [*Hint*: It suffices to show that (a) implies (b), (b) implies (c), and (c) implies (a). That (a) implies (b) is easy. To show (b) implies (c) suppose that U contains no smallest element. Let $T(n)$ be the statement that n is not in U. Use (b) to show U is void. Finally, to show that (c) implies (a) suppose $S(1)$ is true and $S(n)$ implies $S(n + 1)$. Let U be the set of n such that $S(n)$ is false. Show U contains no smallest element.]

(170) *Note.* Statement (a) of the previous exercise is what is usually meant by "mathematical induction," but (b) is often more convenient and will be used in the proposition to follow. Of course the exercise proves none of the statements, only that they are equivalent. None will be proved here. We assume (c) as a basic property of the integers.

(171) **Proposition.** Suppose $\Sigma_{d|n} f(d) = \Sigma_{d|n} g(d)$ for all positive integers n. Then $f = g$.

Proof. Let $T(n)$ be the statement that $f(n) = g(n)$. Since $f(1) = \Sigma_{d|1} f(d) = \Sigma_{d|1} g(d) = g(1)$, $T(1)$ is true. Suppose $T(1)$, $T(2), \ldots$, $T(n)$ true. Let d_1, $d_2, \ldots, d_t$ be the positive divisors of $n + 1$ other than $n + 1$ itself. By the last assumption $f(d_i) = g(d_i)$ for $i = 1, 2, \ldots, t$. Thus

$$f(n + 1) = \sum_{d|n+1} f(d) - \sum_{i=1}^{t} f(d_i) = \sum_{d|n+1} g(d) - \sum_{i=1}^{t} g(d_i) = g(n + 1).$$

This is $T(n + 1)$. Thus $f(n) = g(n)$ for all positive integers n.

(172) Our method of attacking the remaining questions of (165) will be as follows. First we will try to get a formula for f, assuming its existence and continuing along the lines of (166). If we are successful we will then drop the assumption that f exists and *define* f by means of this formula. Then, hopefully, we will prove that when so defined f satisfies $F(n) = \Sigma_{d|n} f(d)$ for all n.

Let us try to extend the table of (167) in a more general way, again assuming $F(n) = \Sigma_{d|n} f(d)$. Since $F(n)$ depends on the divisors of n, it is natural to consider the prime factorization of n. If p is prime, $F(p) = \Sigma_{d|p} f(d) = f(1) + f(p) = F(1) + f(p)$. Thus $f(p) = F(p) - F(1)$. Likewise, $F(p^2) = f(1) + f(p) + f(p^2) = F(1) + F(p) - F(1) + f(p^2)$, or $f(p^2) = F(p^2) - F(p)$. We see that in general $F(p^\alpha) = f(1) + f(p) + \cdots + f(p^{\alpha-1}) + f(p^\alpha) = F(p^{\alpha-1}) + f(p^\alpha)$, so $f(p^\alpha) = F(p^\alpha) - F(p^{\alpha-1})$.

n	$f(n)$
p	$F(p) - F(1)$
p^2	$F(p^2) - F(p)$
p^α	$F(p^\alpha) - F(p^{\alpha-1})$

Note that if n is a power of a prime p, then $f(n) = F(n) - F(n/p)$.

Now let us consider $n = pq$, where p and q are distinct primes. Since the positive divisors of n are $1, p, q$, and pq, we have

$$F(pq) = f(1) + f(p) + f(q) + f(pq)$$
$$= F(1) + F(p) - F(1) + F(q) - F(1) + f(pq),$$

or

$$f(pq) = F(pq) - F(p) - F(q) + F(1).$$

What about $f(p^2q)$? Since $F(p^2) = f(1) + f(p) + f(p^2)$, we have

$$F(p^2q) = f(1) + f(p) + f(p^2) + f(q) + f(pq) + f(p^2q)$$
$$= F(p^2) + F(q) - F(1) + F(pq) - F(p) - F(q) + F(1) + f(p^2q).$$

We see $f(p^2q) = F(p^2q) - F(p^2) - F(pq) + F(p)$.

(173) **Exercise.** Compute $f(p^2q^2)$.

(174) We note that the formula for $f(p^2q)$ differs from that of $f(pq)$ only in having an extra p in the *argument* of each F. Using n in these formulas makes the pattern clearer.

n	$f(n)$
pq	$F(n) - F(n/q) - F(n/p) + F(n/pq)$
p^2q	$F(n) - F(n/q) - F(n/p) + F(n/pq)$

Written this way the formulas look identical. We found before that if $n = p^\alpha$ then $f(n) = F(n) - F(n/p)$ no matter what α is. Perhaps the state of affairs for two primes is similar.

(175) *Conjecture.* If $n = p^\alpha q^\beta$, where p and q are distinct primes and α and β are positive integers, then $f(n) = F(n) - F(n/q) - F(n/p) + F(n/pq)$.

(176) In order to try to prove (175) let us take stock of where we are. We are assuming that $F(n) = \Sigma_{d|n} f(d)$ for all positive n. We are assuming nothing at all about f. Since this function is completely arbitrary the equation which is the conclusion of (175) must, if valid, be an *identity* in the values of f. The equation in question may be written as

$$f(n) = \sum_{d|n} f(d) - \sum_{d|n/q} f(d) - \sum_{d|n/p} f(d) + \sum_{d|n/pq} f(d)$$

$$= \sum_{d|p^\alpha q^\beta} f(d) - \sum_{d|p^\alpha q^{\beta-1}} f(d) - \sum_{d|p^{\alpha-1} q^\beta} f(d) + \sum_{d|p^{\alpha-1} q^{\beta-1}} f(d).$$

Clearly $f(d)$ can appear at all only if $d|n$, and at most once in any given summation. Since $n = p^\alpha q^\beta$ we may assume $d = p^\gamma q^\delta$, $\gamma \leqslant \alpha$, $\delta \leqslant \beta$. The first summation contains $f(d)$ for any γ and δ, but we pick up $f(d)$ in the second summation only if $\delta \leqslant \beta - 1$, in the third only if $\gamma \leqslant \alpha - 1$, and in the fourth only if both these inequalities hold. We must have one and only one of the following four cases:

(a) $\gamma \leqslant \alpha - 1$ and $\delta \leqslant \beta - 1$,
(b) $\gamma \leqslant \alpha - 1$ but $\delta = \beta$,
(c) $\delta \leqslant \beta - 1$ but $\gamma = \alpha$,
(d) $\gamma = \alpha$ and $\delta = \beta$.

Let us make a table listing the occurrences of $f(d)$ in each of these cases.

	$f(n)$	1st Σ	2nd Σ	3rd Σ	4th Σ
(a)		$f(d)$	$-f(d)$	$-f(d)$	$f(d)$
(b)		$f(d)$		$-f(d)$	
(c)			$-f(d)$		$f(d)$
(d)	$f(d)$	$f(d)$	(In this case $d = n$.)		

Thus the equation is an identity in $f(d)$ and the conjecture is proved.

(177) **Exercise.** Prove that if $n = pqr$ where p, q, and r are distinct primes, then

$$f(n) = F(n) - F(n/p) - F(n/q) - F(n/r) + F(n/pq) + F(n/pr) + F(n/qr) - F(n/pqr).$$

(178) **Exercise.** Prove that the formula of the previous exercise holds if $n = p^\alpha q^\beta r^\gamma$, where α, β, and γ are arbitrary positive integers.

18 THE μ FUNCTION

(179) We could now continue with n divisible by four primes, but the reader, especially if he has gone through the last exercise, will probably agree that these calculations have reached a complexity that suggests a stab at a general formula for $f(n)$. We have, assuming (178),

n	$f(n)$
p^α	$F(n) - F\left(\dfrac{n}{p}\right)$
$p^\alpha q^\beta$	$F(n) - F\left(\dfrac{n}{p}\right) - F\left(\dfrac{n}{q}\right) + F\left(\dfrac{n}{pq}\right)$
$p^\alpha q^\beta r^\gamma$	$F(n) - F\left(\dfrac{n}{p}\right) - F\left(\dfrac{n}{q}\right) - F\left(\dfrac{n}{r}\right) + F\left(\dfrac{n}{pq}\right) + F\left(\dfrac{n}{pr}\right) + F\left(\dfrac{n}{qr}\right) - F\left(\dfrac{n}{pqr}\right)$

It appears that in general $f(n)$ is $F(n)$ minus $\Sigma F(n/d)$, where d runs through the distinct primes dividing n, plus $\Sigma F(n/d)$, where d runs through all products of pairs of distinct primes dividing n, minus $\Sigma F(n/d)$, where d runs through all products of triples of distinct primes dividing n, etc.

Let us try to say this better. Whether $F(n/d)$ occurs at all, and, if so, with a plus or minus sign, seems to depend entirely on d, and it appears that we need only worry about d's dividing n. Let us write

$$f(n) = \sum_{d\mid n}{}' \mu(d)F(n/d).$$

and try to define μ so as to express our conjecture. Clearly we want $\mu(d)$ to be 1 if d is a product of an even number of distinct primes and -1 if d is a product of an odd number of distinct primes. We don't want $F(n/d)$ appearing at all if d is not a product of distinct primes, so let us define $\mu(d)$ to be 0 for all other values of d. A special case is $d = 1$; since we want $F(n)$ to appear let us set $\mu(1) = 1$.

(180) **Definition.** We define the function μ on the positive integers by

$$\mu(d) = \begin{cases} 1 \text{ if } d = 1 \\ 0 \text{ if } p^2 \mid d \text{ for some prime } p \\ (-1)^t \text{ if } d \text{ is the product of } t \text{ distinct primes.} \end{cases}$$

(181) **Examples.** $\mu(2) = -1, \mu(6) = 1, \mu(8) = 0, \mu(30) = -1$.

(182) **Exercise.** Compute $\mu(n)$ for $n = 15, 16, 17, 18, 19, 20, 21$.

19 THE MÖBIUS INVERSION FORMULA

(183) Our original problem was to find a function f such that $F(n) = \Sigma_{d \mid n} f(d)$
for all positive integers n, given an arbitrary function F. We think we know what
f must be. Let us assume F is given and *define* f by $f(n) = \Sigma_{d \mid n} \mu(d) F(n/d)$,
$n \geqslant 1$. We must now show that if we sum $f(n)$ as n runs through the positive divisors of some other positive integer, say k, then we get $F(k)$. That is, we must
prove that for any positive integer k,

(*)
$$\sum_{n \mid k} \left(\sum_{d \mid n} \mu(d) F(n/d) \right) = F(k).$$

This equation shows how the Σ notation may be used to write in a compact way
something which would be almost impossible to express with +'s and . . . 's.
Imagine trying to write (*) without using Σ's. Now imagine trying to write (*)
without using Σ's so that someone else could understand what was meant. On
the other hand, complicated combinations of Σ's and Π's have a tendency to be-
come a sort of independent mumbo jumbo, their meaning as sums and products
forgotten. This is dangerous. The reader coming upon an expression such as (*)
should stop and make sure he knows what it really means. Let us break (*)
down partially to get a clearer idea of what it says. Suppose $n_1, n_2, \ldots, n_t$ are
the positive divisors of k. Then we can rewrite (*) as

(**)
$$F(k) = \sum_{d \mid n_1} \mu(d) F(n_1/d) + \sum_{d \mid n_2} \mu(d) F(n_2/d) + \cdots$$
$$+ \sum_{d \mid n_t} \mu(d) F(n_t/d).$$

Since F is an arbitrary function we must show that this is an identity in its
values. Let a be any positive integer. We must show that either $F(a)$ cancels out
on the right-hand side of (**) (if $a \neq k$), or else appears there with a net coefficient of +1 (if $a = k$). Let us consider this last case first. For $F(a)$ to appear
on the right-hand side of (**) we must have $a = n/d$, where $n \mid k$ and $d \mid n$. If $a = k$
then clearly $n = k$ and $d = 1$. Thus $F(k)$ appears exactly once in the right-hand

side, namely in the $d = 1$ term of the summation where $n_i = k$. Since $\mu(d) = \mu(1) = 1$, (**) checks out for $F(k)$.

   Now let us suppose $a \neq k$. First we note that $F(a)$ can appear at most once in any one summation on the right-hand side of (**). This is because n/d is different for different values of d. As before, for $F(a)$ to appear at all we must have $a = n/d$, where $n|k$ and $d|n$. We see $a|k$. For each combination of positive integers d and n such that $n|k$, $d|n$, and $n/d = a$, we pick up an $F(a)$, with coefficient $\mu(d)$. Thus we must show that

$$\sum_{\substack{n|k, d|n \\ a = n/d}} \mu(d) = 0.$$

This appears to be a double summation, with variables n and d, but in reality because of the relation $n/d = a$ once we choose one of these variables the other is determined. Since $n = ad$, we could rewrite what we want as $\sum_{\substack{da|k \\ d|da}} \mu(d) = 0$.

But $d|da$ is always true, so we can simplify this to $\sum_{da|k} \mu(d) = 0$.

(184) **Exercise.** What is the variable of summation in the last expression? If you don't know, your eyes may be moving but you're not reading.

(185) Here's where we are. We are trying to prove that $\sum_{n|k} (\sum_{d|n} \mu(d) F(n/d)) = F(k)$ is an identity in $F(a)$. We have noted that we can assume $a|k$ and have taken care of the case $a = k$. We have shown moreover that the net coefficient of $F(a)$ is $\sum_{da|k} \mu(d)$. It only remains to show that this is 0 for $a < k$. For example, let $k = 24$ and $a = 2$. Then

$$\sum_{2d|24} \mu(d) = \mu(1) + \mu(2) + \mu(3) + \mu(4) + \mu(6) + \mu(12)$$
$$= \quad 1 \quad -1 \quad -1 \quad +0 \quad +1 \quad +0 = 0.$$

In this case we see that the positive integers d such that $2d|24$ are just the positive divisors of 12. We wonder whether $ad|k$ is equivalent to $d|k/a$ in general. This is easy to see; $ad|k$ if and only if $k/ad = (k/a)/d$ is an integer, i.e., if and only if $d|k/a$. We can write our equation as $\sum_{d|k/a} \mu(d) = 0$. Since the only restrictions on k and a are that $a|k$ and $a < k$, this amounts to

(186) **Proposition.** If $m > 1$, then $\sum_{d|m} \mu(d) = 0$.

(187) We have come down to evaluating the function $g(m) = \sum_{d|m} \mu(d)$. This would be fairly easy if we knew g was multiplicative. By (140) it suffices that μ be multiplicative.

(188) **Exercise*.** Prove the function μ is multiplicative.

(189) Let us consider $g(p^\alpha)$, where p is a prime and $\alpha \geqslant 1$. By definition $g(p^\alpha) = \mu(1) + \mu(p) + \mu(p^2) + \cdots + \mu(p^\alpha) = \mu(1) + \mu(p) + 0 + \cdots + 0 = 1 - 1 = 0$.

Since if $m > 1$ then some prime, say p, divides m, we have, assuming $m = p^\alpha b$ where $p \nmid b$, $\Sigma_{d|m} \mu(d) = g(m) = g(p^\alpha b) = g(p^\alpha)g(b) = 0$. Thus we have proved (186), establishing (*). Combining this with (171) we get the following theorem, which answers all three of our original questions.

(190) **THEOREM.** Suppose F is any function defined on the positive integers. Define f by $f(n) = \Sigma_{d|n} \mu(d)F(n/d)$ for all positive integers n. Then $\Sigma_{d|n} f(d) = F(n)$ for all n. Furthermore, f is the only function having this property.

(191) *Note.* The formula of the preceding theorem is called the *Möbius Inversion Formula* and μ is called the *Möbius function.*

(192) *Note.* Suppose we know $F(n) = \Sigma_{d|n} f(d)$ for all n. Define g by $g(n) = \Sigma_{d|n} \mu(d)F(n/d)$. Then $F(n) = \Sigma_{d|n} g(d)$. By the last sentence of (190) we must have $f = g$. Thus $f(n) = \Sigma_{d|n} \mu(d)F(n/d)$.

(193) **Exercise.** Suppose $F(n) = \Sigma_{d|n} f(d)$ for $n \geqslant 1$. Show $f(n) = \Sigma_{d|n} \mu(n/d) \cdot F(d)$ for $n \geqslant 1$.

(194) **Definition.** Suppose f and g are functions and k and c are real numbers. We define the functions $f + g$, fg, f^k, cf, f/g, and $f \circ g$ by $(f + g)(n) = f(n) + g(n)$, $(fg)(n) = f(n)g(n)$, $f^k(n) = (f(n))^k$, $(cf)(n) = cf(n)$, $(f/g)(n) = f(n)/g(n)$, and $(f \circ g)(n) = f(g(n))$ for all n.

(195) **True-False.** Assume f and g are multiplicative functions, with $g(n) \neq 0$ for all n, and let k and c be constants, $k > 0$.

   (a) $f + g$ is multiplicative.
   (b) fg is multiplicative. ·
   (c) cf is multiplicative.
   (d) f^k is multiplicative.
   (e) f/g is multiplicative.
   (f) $f \circ g$ is multiplicative.
   (g) $f(1) = 1$.
   (h) $g(1) = 1$.

20 EVALUATING MULTIPLICATIVE FUNCTIONS

(196) The reader was not seeing the technique we used to evaluate $\Sigma_{d|n} \mu(d)$ in (187) through (189) for the first time, nor, for that matter, for the last. (It will be very useful in the exercises to follow.) We can summarize it as follows:

   To evaluate $F(n) = \Sigma_{d|n} f(d)$
   1. Prove f is multiplicative
   2. Conclude F is multiplicative by (140)
   3. Evaluate $F(p^\alpha)$, p prime
   4. Evaluate $F(n)$ in general by means of (2) and (3).

(Of course if f is *not* multiplicative the whole thing fails and we must look for another method.)

A good question is whether this argument can be turned around. For example, suppose we were given that $\Sigma_{d|n}\,\mu(d) = \begin{cases} 1 \text{ if } n = 1 \\ 0 \text{ if } n > 1 \end{cases}$. This is easily seen to by multiplicative; does it follow that μ is? Is it true in general that if F is multiplicative then so is f? This would make (140) into an "if and only if" theorem. Let us suppose $F(n) = \Sigma_{d|n}\,f(n)$ is multiplicative. It seems natural to use (192) to write f in terms of F. Suppose $(a,b) = 1$. Then

$$f(a)f(b) = \left(\sum_{d|a} \mu(d)F(a/d)\right)\left(\sum_{e|b} \mu(e)F(b/e)\right)$$

$$= \sum_{\substack{d|a \\ e|b}} \mu(d)\mu(e)F(a/d)F(b/e).$$

(If this last step seems mysterious to the reader he should let $d_1, d_2, \ldots, d_s$ and $e_1, e_2, \ldots, e_t$ be the divisors of a and b respectively and write out the sums involved. It is an application of (138) with $g(d) = \mu(d)F(a/d)$ and $h(e) = \mu(e)F(b/e)$.) Since no prime divides both a and b the same can be said about d and e; so $(d,e) = 1$. Likewise $(a/d,b/e) = 1$. Thus we have $f(a)f(b) = \Sigma_{\substack{d|a \\ e|b}} \mu(de)F(ab/de)$, where we have used the fact that both μ and F are multiplicative. We could say that this was $f(ab) = \Sigma_{h|ab}\,\mu(h)F(ab/h)$ if we knew that as d and e ran through the positive divisors of a and b, de ran through the positive divisors of ab, without repetition. But we know exactly this; it is (120). Combining this result with (140) we have

(197)   **THEOREM.** Suppose $F(n) = \Sigma_{d|n}\,f(d)$ for $n \geqslant 1$. Then F is multiplicative if and only if f is.

(198)   **Exercise.** Prove $\Sigma_{d|n}\,\mu(d)\tau(n/d) = 1$ for $n \geqslant 1$.

(199)   **Exercise.** Prove $\Sigma_{d|n}\,\mu(d)\tau(d) = \Pi_{\substack{p|n \\ p \text{ prime}}} (-1), n \geqslant 1$.

(200)   **Exercise.** Prove $\Sigma_{d|n}\,\mu(d)\sigma(d) = \Pi_{\substack{p|n \\ p \text{ prime}}} (-p), n \geqslant 1$.

(201)   **Exercise.** If $F(n) = \Sigma_{d|n}\,f(d)$ for $n \geqslant 1$ we write $F = f'$. Prove that $\mu''' = \tau$.

(202)   **Exercise.** Let $m(n) = 0$ if $p^3|n, p$ prime. Otherwise let

$$m(n) = \Pi_{\substack{p||n \\ p \text{ prime}}} (-2).$$

Prove that $m' = \mu$ (notation of (201)).

chapter 3

The Arithmetic of Congruence Classes

A fixed modulus induces a natural decomposition of the integers into disjoint subsets of congruent numbers. Various facts about the resulting structure will be uncovered in this chapter in the course of proving that a famous function of L. Euler is multiplicative.

21 THE EULER φ FUNCTION

(203) There has hardly been anything we have done up to now in which the prime numbers have not played an important role. Although being prime is an all-or-nothing proposition, we have the feeling that some numbers are more composite than others. For example $60 = 2 \cdot 2 \cdot 3 \cdot 5$ seems further from primeness than $62 = 2 \cdot 31$. One way we might quantify this feeling is with $\tau(n)$, the number of positive divisors of n. Only if n is prime can $\tau(n)$ be 2; for composite numbers it is bigger. In a certain sense the bigger $\tau(n)$ is the further n is from being prime. For example $\tau(62) = 4$, while $\tau(60) = 12$.

Another way to measure the same thing depends on the fact that if p is prime, then $(p,n) > 1$ if and only if $p \mid n$. This property characterizes the primes, for if a is composite there exists an n such that $n \mid a$ and $1 < n < a$. Then $(a,n) > 1$ but $a \nmid n$. Other things being equal, we expect $(a,n) = 1$ to be a more common occurrence when a is prime than when a is composite. As an example let us write out the first few n such that $(7,n) = 1$. The easiest way is to start writing out all the integers, crossing out the multiples of 7.

$$
\begin{array}{ccccccc}
1 & 2 & 3 & 4 & 5 & 6 & \cancel{7} \\
8 & 9 & 10 & 11 & 12 & 13 & \cancel{14} \\
15 & 16 & 17 & 18 & 19 & 20 & \cancel{21}
\end{array}
$$

Let us now do the same thing with 6 instead of 7. Since $(a,6) > 1$ if and only if some prime divides both a and $6 = 2 \cdot 3$, it suffices to cross out the multiples of 2 and 3.

$$1 \quad \cancel{2} \quad \cancel{3} \quad \cancel{4} \quad 5 \quad \cancel{6}$$
$$7 \quad \cancel{8} \quad \cancel{9} \quad \cancel{10} \quad 11 \quad \cancel{12}$$
$$13 \quad \cancel{14} \quad \cancel{15} \quad \cancel{16} \quad 17 \quad \cancel{18}$$

The n such that $(6, n) = 1$ appear to be much rarer.

The reader has probably noticed a pattern in the two arrays just presented. Either all or none of the numbers in each column have been crossed out. This is no surprise if we recall that we proved in (66) that if $b \equiv b'$ (mod a), then $(a, b) = (a, b')$. In particular, $(a, b) = 1$ if and only if $(a, b') = 1$. In the first array we listed the numbers in 7 columns, so those in any given column were all congruent modulo 7. In the second array the columns consisted of numbers congruent modulo 6. Since we cannot count all the n such that $(a, n) = 1$, there being infinitely many of them, let us merely count those in the "first row"; after that the pattern repeats anyway.

(204) **Definition.** Suppose a is a positive integer. We define $\varphi(a)$ to be the number of integers n, $1 \leqslant n \leqslant a$, such that $(a, n) = 1$. The function φ is called the *Euler phi-function.*

(205) By what we have already seen, $\varphi(6) = 2$ and $\varphi(7) = 6$. Clearly $\varphi(p) = p - 1$ for any prime p. By definition $\varphi(1) = 1$. Let us compute $\varphi(12)$. Since $12 = 2^2 \cdot 3$, it suffices to cross out the multiples of 2 and 3 from among the first 12 integers:

$$1 \quad \cancel{2} \quad \cancel{3} \quad \cancel{4} \quad 5 \quad \cancel{6} \quad 7 \quad \cancel{8} \quad \cancel{9} \quad \cancel{10} \quad 11 \quad \cancel{12}.$$

Thus $\varphi(12) = 4$.

(206) **Exercise.** Compute $\varphi(n)$ for $n = 4, 18, 19$, and 30.

(207) Computing $\varphi(12)$ by our "crossing-out" method amounted to eliminating the same numbers as in computing $\varphi(6)$, the multiples of 2 and 3. Thus the pattern from 7 through 12 repeats that of 1 through 6. If we had noticed this at the start we could have cut our work in half. We could have crossed out the multiples of 2 and 3 among the first 6 integers, leaving 2 numbers, then doubled this count of 2 to get $\varphi(12) = 4$. In other words, $\varphi(12) = 2\varphi(6)$. In the same way $\varphi(18) = 3\varphi(6) = 6$. This works since 18 has no prime divisor that does not already divide 6. Note that $\varphi(30) = 8 \neq 5\varphi(6)$.

(208) **Proposition.** If any prime dividing k also divides a, then $\varphi(ka) = k\varphi(a)$.
    Proof. Consider the array

$$
\begin{array}{llll}
1, & 2, & \ldots, & a, \\
a + 1, & a + 2, & \ldots, & 2a, \\
& \cdots & & \\
(k - 1)a + 1, & & \ldots, & ka.
\end{array}
$$

Let us cross out the entries which are not relatively prime to ka in order to compute $\varphi(ka)$. This amounts to crossing out everything not relatively prime to a, since $(a, n) > 1$ if and only if some prime divides both a and n; and the same primes divide ka as divide a. By (66) the pattern of crossings-out is the same in each row. Since there are $\varphi(a)$ entries left in the first row, there must be $k\varphi(a)$ entries left in all. Thus $\varphi(ka) = k\varphi(a)$.

(209) **Exercise.** Compute $\varphi(n)$ for $n = 36, 108, 128$, and 243.

22   A FORMULA FOR φ

(210) **Proposition** (208) goes a long way toward a formula for $\varphi(n)$. For example, suppose we want $\varphi(10^6)$. We have $\varphi(10^6) = \varphi(2^6 \cdot 5^6) = \varphi(2^5 \cdot 5^5 \cdot 2 \cdot 5) = 2^5 \cdot 5^5 \, \varphi(10) = 100,000\varphi(10)$, but we still must compute $\varphi(10)$ by an actual count. The computation of $\varphi(n)$ can always be reduced in the same way to that of φ of a product of distinct primes. If we could pull the p_1 out of $\varphi(p_1 \cdot p_2 \cdots p_t)$ just as we learned to pull the k out of $\varphi(ka)$ under the hypothesis of (208) we could write a complete formula. Let us see if we can modify the proof of (208) to compute $\varphi(pa)$, where p is a prime *not* dividing a. As before, let us write out the integers from 1 to pa

$$
\begin{array}{cccc}
1, & 2, \ldots, & a, \\
a + 1, & & \ldots, 2a, \\
& \cdots & \\
(p - 1)a + 1, & & \ldots, pa.
\end{array}
$$

Again let us imagine crossing out the n such that $(pa, n) > 1$. The difference here is that $(pa, n) > 1$ is not equivalent to $(a, n) > 1$; the latter implies the former but not conversely. Thus of the $p\varphi(a)$ integers left after crossing out all n such that $(a, n) > 1$, more still must be eliminated, namely the multiples of p. These are just $1 \cdot p, 2p, \ldots, ap$. Of these, those kp such that $(k, a) > 1$ have already been eliminated; there are just $\varphi(a)$ more to cross out in order that $\varphi(pa)$ be properly counted. Thus

$$\varphi(pa) = p\varphi(a) - \varphi(a) = (p - 1)\varphi(a).$$

(211) **Examples.** $\varphi(10^6) = 10^5 \varphi(2 \cdot 5) = 10^5 (2 - 1)\varphi(5) = 400000$. In the same way $\varphi(60) = \varphi(2^2 \cdot 3 \cdot 5) = 2\varphi(2 \cdot 3 \cdot 5) = 2(2 - 1)\varphi(3 \cdot 5) = 2(3 - 1)\varphi(5) = 16$ and $\varphi(62) = \varphi(2 \cdot 31) = (2 - 1)\varphi(31) = 30$.

(212) **Exercise.** Compute $\varphi(n)$ for $n = 117, 152$, and 1001.

(213) In general if $n = p_1^{\alpha_1} p_2^{\alpha_2} \ldots p_t^{\alpha_t}$, where the p's are distinct primes, then $\varphi(n) = p_1^{\alpha_1 - 1} (p_1 p_2^{\alpha_2} \ldots p_t^{\alpha_t}) = p_1^{\alpha_1 - 1} (p_1 - 1) \, \varphi(p_2^{\alpha_2} \ldots p_t^{\alpha_t}) = \cdots = p_1^{\alpha_1 - 1}) \cdot (p_1 - 1) \ldots p_t^{\alpha_t - 1} (p_t - 1) = p_1^{\alpha_1} p_2^{\alpha_2} \ldots p_t^{\alpha_t} (p_1 - 1) (p_2 - 1) \ldots (p_t - 1) / p_1 p_2 \ldots p_t = n(1 - 1/p_1)(1 - 1/p_2) \ldots (1 - 1/p_t)$.

(214) **THEOREM.** $\displaystyle \varphi(n) = \prod_{\substack{p^\alpha \| n \\ p \text{ prime}}} p^{\alpha - 1} (p - 1) = n \prod_{\substack{p \mid n \\ p \text{ prime}}} \left(1 - \frac{1}{p}\right).$

(215) **Examples.** Using the first formula $\varphi(10^6) = \varphi(2^6 \cdot 5^6) = 2^5 (2 - 1) 5^5 \cdot (5 - 1); \varphi(360) = \varphi(2^3 3^2 5) = 2^2 (2 - 1) 3 (3 - 1) (5 - 1) = 96$. Using the second formula $\varphi(10^6) = \varphi(2^6 5^6) = 10^6 (1 - \frac{1}{2}) (1 - \frac{1}{5}); \varphi(360) = \varphi(2^3 3^2 5) = 360 \cdot (1 - \frac{1}{2}) (1 - \frac{1}{3}) (1 - \frac{1}{5}) = 360 (\frac{1}{2}) (\frac{2}{3}) (\frac{4}{5}) = 96$.

(216) **Exercise.** Compute $\varphi(n)$ for $n = 17, 32, 34, 40, 48$, and 60.

(217) **Exercise.** Prove that $\varphi(n) \longrightarrow \infty$ as $n \longrightarrow \infty$; i.e., prove given M there exists n_0 such that $n \geqslant n_0$ implies $\varphi(n) \geqslant M$.

(218) **Exercise.** Find n_0 such that $n \geqslant n_0$ implies $\varphi(n) \geqslant 100$.

(219) **Exercise.** Show that for n fixed $\varphi(x) = n$ has at most a finite number of solutions.

(220) **Exercise.** Find all solutions to $\varphi(x) = n$ for $n = 1, 2$, and 4. Prove you have found them all.

(221) **Exercise.** Find all solutions to $\varphi(x) = 14$. Prove you have found them all.

(222) **Exercise*.** Use (214) to prove φ is multiplicative.

(223) **Exercise.** Prove that $\Sigma_{d|n} \varphi(d) = n$ for $n \geqslant 1$.

(224) **Exercise.** Prove $\varphi'' = \sigma$, in the notation of (201).

(225) **Exercise.** Let $l(n) = n \prod_{\substack{p^2 | n \\ p \text{ prime}}} \left(1 - \frac{1}{p}\right)^2 \prod_{\substack{p \| n \\ p \text{ prime}}} \left(1 - \frac{2}{p}\right)$. Prove $l' = \varphi$.

(226) **Exercise.** Prove $\Sigma_{d|n} \mu(d) \varphi(d) = \Pi_{\substack{p|n \\ p \text{ prime}}} (2 - p)$ for $n \geqslant 1$.

(227) **Exercise.** Let $\epsilon > 0$. Prove there exists n such that $|\varphi(n)/n - 1| < \epsilon$.

(228) **Exercise*.** Prove directly from the definition of φ that $\varphi(p^\alpha) = p^{\alpha-1} (p - 1)$ for p prime.

(229) **Exercise.** Determine all n such that $\varphi(n)$ is odd.

(230) **Exercise.** Find all solutions to $\varphi(x) = \varphi(2x)$.

(231) **Exercise.** Find all solutions to $\varphi(2x) = \varphi(3x)$.

23 PROVING φ MULTIPLICATIVE

(232) Our proof of the formula for $\varphi(n)$ was a switch. If we had followed our standard manner of attacking such problems we would have first proved φ multiplicative, then derived the formula for $\varphi(n)$ from that of $\varphi(p^\alpha)$, p prime. This would have worked; exercises (222) and (228) show that φ *is* multiplicative and that $\varphi(p^\alpha)$ can be evaluated directly from the definition of φ by a simple count. Of course we would need an *independent* proof that φ was multiplicative. Let us attempt to find such a proof.

(233) We showed that τ was multiplicative by proving that as d and e ran through the positive divisors of a and b respectively, de ran through the positive divisors of ab, assuming $(a,b) = 1$. Perhaps a similar argument will work for φ. Again we assume $(a,b) = 1$. Let the positive integers not exceeding a and relatively prime to a be $n_1, \ldots, n_s$. Thus $s = \varphi(a)$. Let the corresponding integers for b be $m_1, \ldots, m_t$, where $t = \varphi(b)$. Hopefully the integers $n_i m_j$, $1 \leqslant i \leqslant s$, $1 \leqslant j \leqslant t$, will comprise exactly the positive integers relatively prime to ab and not exceeding ab.

As an example, let us take $a = 5$ and $b = 6$. Then the n's are 1, 2, 3, and 4 and the m's are 1 and 5. The nm's (which we hope are the positive integers not exceeding 30 and prime to 30) are 1, 2, 3, 4, 5, 10, 15, and 20. Disaster! Of these numbers only the first, 1, is prime to 30. One hit and seven misses is not a very good average.

We can see what is going wrong. We have $(n,5) = 1$ and $(m,6) = 1$ and would like $(nm,30) = 1$. But for nm to be prime to 30 we clearly require that both n and m be. Having $(n,5) = 1$ is not enough; we need $(n,6) = 1$ also. Likewise for the m's. Our m's were 1 and 5. But $a = 5$, so we cannot expect $(n \cdot 5, 30) = 1$ for any n.

This seems to be partly just bad luck. The m's were 1 and 5, and it just so happened that $a = 5$. If we had chosen to do an example with $a = 4$ or $a = 7$, at least this embarrassment would have been avoided. (Though there would still be trouble with the n's.)

(234) Of course we want an argument that works for *any* a and b. The point of the discussion above is that difficulties arising from "accidental" properties of a and b can perhaps be gotten around. Let us recall what the n's and m's are. We continue with the example $a = 5$, $b = 6$. The m's, 1 and 5, are the uncrossed-out elements of the first row of the array

$$
\begin{array}{ccccccc}
① & \not{2} & \not{3} & \not{4} & ⑤ & \not{6} \\
7 & \not{8} & \not{9} & \not{10} & 11 & \not{12} \\
13 & \not{14} & \not{15} & \not{16} & 17 & \not{18} \\
19 & \not{20} & \cdots
\end{array}
$$

As such, they merely *count* the columns of integers prime to 6. There is nothing sacred about the first row; any other could have been used just as well. All that is really needed to count $\varphi(b)$, in fact, is that exactly one element be chosen from each column of numbers prime to b.

24 REDUCED AND COMPLETE RESIDUE SYSTEMS

(235) **Definition.** Let b be any positive integer. We say that $m_1, m_2, \ldots, m_t$ is a *reduced residue system modulo* b in case $(m_i, b) = 1$ for $i = 1, 2, \ldots, t$, and given any integer r relatively prime to b there exists exactly one i, $1 \leqslant i \leqslant t$, such that $m_i \equiv r \pmod{b}$.

(236) **Examples.** The following are reduced residue systems modulo 6: 1,5; 7,11; 1,11; -1,1; -1,-5; 601,605. The following are not: 1,4; 1,4,5; -1,5; 1,1,5; 11; 1,-1,5. The numbers 1,-1,2,-2 form a reduced residue system modulo 5; so do the numbers 3, 6, 9, 12; the numbers 4, 7, 10, 13 do not.

(237) *Note.* In (235) of necessity $t = \varphi(b)$. That's the idea.

(238) **Definition.** We say $m_1, m_2, \ldots, m_b$ is a *complete residue system modulo* b in case given any integer r there exists exactly one i, $1 \leqslant i \leqslant b$, such that $m_i \equiv r \pmod{b}$.

(239) **Examples.** The following are complete residue systems modulo 6: 1, 2, 3, 4, 5, 6; 0, 1, 2, 3, 4, 5; 7, 14, 21, 28, 35, 42; -60, 61, -58, 63, -56, 65. The following are not: 2, 4, 6, 8, 10, 12; 1, -1, 2, -2, 3, -3; 0, 1, 2, 3, 4, 5, 6; 1, 2, 3, 4, 5; 1, 4, 9, 2, 4, -6.

(240) **Definition.** Let b be a positive integer. Any set of the form $\{x : x \equiv r \pmod{b})\}$, r fixed, is called a *congruence class modulo* b.

(241) **Example.** One congruence class modulo 6 is $\{3, -3, 9, -9, 15, -15, \ldots\}$. Another is $\{\ldots, -11, -5, 1, 7, 13, \ldots\}$. The odd integers form one congruence class modulo 2, the even integers another.

(242) **Exercise*.** Prove that any two congruence classes modulo b are either disjoint or identical.

(243) *Note.* The congruence classes modulo b are just the "columns" referred to above (including negative numbers). There are exactly b of them.

(244) **True-False.** Suppose that b and r are integers, $b > 0$. Suppose m_1, $m_2, \ldots, m_t$ is a reduced residue system modulo b and $m_1, m_2, \ldots, m_b$ is a complete residue system modulo b.

(a) $m_1 + r, m_2 + r, \ldots, m_t + r$ is a reduced residue system modulo b.
(b) $m_1 + r, \ldots, m_b + r$ is a complete residue system modulo b.
(c) If $m_1 + r, \ldots, m_t + r$ is a reduced residue system modulo b, then $b|r$.
(d) $-m_1, -m_2, \ldots, -m_t$ is a reduced residue system modulo b.
(e) $-m_1, \ldots, -m_b$ is a complete residue system modulo b.
(f) There exists a complete residue system modulo b, say $k_1, \ldots, k_b$, such that $|k_i| \leqslant b/2, i = 1, 2, \ldots, b$.
(g) There exists a complete residue system modulo b, say $k_1, \ldots, k_b$, such that $|k_i| < b/2, i = 1, 2, \ldots, b$.
(h) There exists a reduced residue system modulo b, say $k_1, k_2, \ldots, k_t$, such that $|k_i| < b/2, i = 1, 2, \ldots, t$.
(i) $m_1^2, m_2^2, \ldots, m_b^2$ is a complete residue system modulo b.
(j) $m_1^2, m_2^2, \ldots, m_t^2$ is a reduced residue system modulo b.

25 MULTIPLES OF RESIDUE SYSTEMS

(245) Let us go back to the problem of showing that φ is multiplicative. It would suffice to show that as n ran through a reduced residue system modulo a and m ran through a reduced residue system modulo b, then nm ran through a reduced residue system modulo ab, since the number of nm's is clearly $\varphi(a)\varphi(b)$. We hope that using reduced residue systems will give us enough freedom in choosing the n's and m's to avoid the difficulty that arose in (233). Using the example $a = 5$, $b = 6$ again, we want n's and m's prime to *both* 5 and 6. Consider the table for 6 again.

①	2	3	4	5	6
7	8	9	10	⑪	12
13	14	. . .			

Keeping in mind that $a = 5$ we choose the reduced residue system 1, 11 instead of 1, 5.

The table for 5 is

①	2	3	4	5
6	⑦	8	9	10
11	12	⑬	14	15
16	17	18	⑲	20
21	. . .			

Here we choose the reduced residue system 1, 7, 13, 19, the elements of which are all prime to 6. Thus we have

$$n\text{'s: } 1, 11$$
$$m\text{'s: } 1, 7, 13, 19$$
$$nm\text{'s: } 1, 7, 13, 19, 11, 77, 143, 209.$$

To check whether the nm's are a reduced residue system modulo 30 let us replace them by their remainders upon division by 30 [see (65)]. We get

$$1, 7, 13, 19, 11, 17, 23, 29.$$

These are distinct; thus the nm's all belong to different congruence classes modulo 30. Since there are the proper number of them $(\varphi(30) = (2 - 1)(3 - 1) \cdot (5 - 1) = 8)$, they must form a reduced residue system modulo 30. (Of course our eventual goal is to prove $\varphi(30) = \varphi(5)\varphi(6)$ without using the formula; we use it here just to see if we are on the right track.)

As another example, let us take $a = 3$ and $b = 4$. We choose reduced residue systems modulo 3 and 4, restricting ourselves to numbers prime to 12.

$$\begin{array}{cccc} \textcircled{1} & 2 & \cancel{3} & \\ 4 & \textcircled{5} & \cancel{6} & \\ 7 & \ldots & & \end{array} \qquad \begin{array}{cccc} \textcircled{1} & \cancel{2} & 3 & \cancel{4} \\ 5 & \cancel{6} & \textcircled{7} & \cancel{8} \\ 9 & \ldots & & \end{array}$$

Thus the *n*'s are 1 and 5, the *m*'s are 1 and 7, and the *nm*'s are 1, 5, 7, and 35.

$$\begin{array}{cccccccccccc} \textcircled{1} & \cancel{2} & \cancel{3} & 4 & \textcircled{5} & \cancel{6} & \textcircled{7} & \cancel{8} & \cancel{9} & \cancel{10} & 11 & \cancel{12} \\ 13 & \cancel{14} & \cancel{15} & 16 & 17 & \cancel{18} & 19 & \cancel{20} & \cancel{21} & \cancel{22} & 23 & \cancel{24} \\ 25 & \cancel{26} & \cancel{27} & 28 & 29 & \cancel{30} & 31 & \cancel{32} & \cancel{33} & 34 & \textcircled{35} & \cancel{36} \end{array}$$

We see that the *nm*'s are a reduced residue system modulo 12.

(246) **Exercise.** Choose reduced residue systems modulo 3 and 10, using the least positive integers prime to 30. Show that the set of all products *nm*, where *n* is in the reduced residue system modulo 3 and *m* is in the reduced residue system modulo 10, is a reduced residue system modulo 30.

(247) Our examples are very promising. Before we can work up a general proof along these lines, however, we must answer the following question. Given relatively prime integers a and b, can we always find reduced residue systems modulo a and b consisting entirely of numbers prime to ab? To be more specific, given n such that $(a,n) = 1$, can we always find n' such that $n' \equiv n$ (mod a) and $(n',b) = 1$? Of course the condition $n' \equiv n$ (mod a) just means that n' is in the same column as n in our table for a, or that $n' = n + ka$ for some integer k. The question is whether we can find an integer k such that $n + ka$ is prime to b.

For example, if $n = 1$, $a = 5$, $b = 6$, and $k = 0, 1, 2, \ldots$, the numbers $n + ka = 1 + 5k$ are $1, 6, 11, 16, \ldots$

$$\begin{array}{cccccc} \textcircled{1} & \cancel{2} & \cancel{3} & \cancel{4} & 5 & \textcircled{6} \\ 7 & \cancel{8} & \cancel{9} & \cancel{10} & \textcircled{11} & \cancel{12} \\ 13 & \cancel{14} & \cancel{15} & \textcircled{16} & 17 & \cancel{18} \\ 19 & \cancel{20} & \textcircled{21} & \cancel{22} & 23 & \cancel{24} \\ 25 & \textcircled{26} & \cancel{27} & \cancel{28} & 29 & \cancel{30} \\ \textcircled{31} & \cancel{32} & \ldots & & & \end{array}$$

In our example the numbers $n + ka$ fall into *every* congruence class modulo b, which is more than we really asked for.

Let us try $a = 6$, $b = 5$, and $n = 1$. We circle the numbers of the form $n + ka = 1 + 6k$ in the 5-table:

$$\begin{array}{ccccc}
① & 2 & 3 & 4 & \cancel{5} \\
6 & ⑦ & 8 & 9 & \cancel{10} \\
11 & 12 & ⑬ & 14 & \cancel{15} \\
16 & 17 & 18 & ⑲ & \cancel{20} \\
21 & 22 & 23 & 24 & \cancel{㉕} \\
26 & 27 & \cdots &&
\end{array}$$

Again the numbers fall into all congruence classes. In both our examples not only does $n + ka$ hit all congruence classes modulo b, but it suffices to consider only $k = 0, 1, 2, \ldots, b - 1$.

We naturally wonder whether in general the numbers $n + ka$ hit all congruence classes modulo b as $k = 0, 1, \ldots, b - 1$. Since we are dealing with exactly b integers, it would suffice to show that they all fall into distinct congruence classes. Let us suppose $n + ka \equiv n + k'a \pmod b$, where $0 \leqslant k, k' \leqslant b - 1$. Then b divides $(n + ka) - (n + k'a) = a(k - k')$. Since $(a,b) = 1$ we must have $b \mid k - k'$. [See (249).] But $|k - k'| < b$, so $k = k'$. We have proved

(248) **Proposition.** Suppose $(a,b) = 1$ and n is any integer. Then the integers $n + ka, k = 0, 1, \ldots, b - 1$, form a complete residue system modulo b.

(249) **Exercise*.** Suppose $(a, b) = 1$ and $b \mid ac$. Prove $b \mid c$. [*Hint:* Consider the prime factorization of ac.]

(250) *Note.* If $k' \equiv k \pmod b$, then $n + k'a \equiv n + ka \pmod b$ by (102). Thus restricting k to the integers $0, 1, \ldots, b - 1$ was too severe; any complete residue system modulo b would have been good enough. In particular, taking $n = 0$, as k runs through any complete residue system modulo b so does ka. If k is prime to b, so is ak, by (80). Thus as k runs through a *reduced* residue system modulo b the $\varphi(b)$ integers ka must also form a reduced residue system modulo b; we already know they are all incongruent [see (259)].

(251) **THEOREM.** Suppose $(a, b) = 1$. As k runs through a complete or reduced residue system modulo b, so does ka.

(252) **Exercise.** Confirm that 0, 9, 18, 27, 36, 45, 54 is a complete residue system modulo 7. Confirm that 7, 14, 28, 35, 49, 56 is a reduced residue system modulo 9.

(253) **Exercise.** Find a reduced residue system modulo 8 consisting of multiples of 9.

(254) **Exercise.** Solve $8x \equiv 3 \pmod 9$, where $1 \leqslant x \leqslant 9$.

(255) **Exercise**. Let $S = \{-4, -2, 0, 2, 8, 12, 19, 22, 24\}$. Show that S is a complete residue system modulo 9. Solve $8x \equiv 3 \pmod 9$, x in S. Solve $8x \equiv 4 \pmod 9$, x in S.

(256) **Exercise**. Find all solutions x of the following, $1 \leqslant x \leqslant 6$.
(a) $5x \equiv 2 \pmod 6$
(b) $2x \equiv 4 \pmod 6$
(c) $3x \equiv 4 \pmod 6$.

(257) **Exercise**. Show that $x = 3n$ is always a solution to $5x \equiv n \pmod 7$.

(258) **Exercise**. Find a general solution (as in (257)) to $7x \equiv n \pmod{16}$.

(259) **Exercise.*** Suppose there are s chickens and t cages. Show that any three of the following conditions implies the fourth.
(a) Each chicken is in some cage.
(b) No two chickens are in the same cage.
(c) Each cage contains at least one chicken.
(d) $s = t$.

26 SIMULTANEOUS CONGRUENCES

(260) We return to our attempt to prove $\varphi(ab) = \varphi(a)\varphi(b)$ when $(a,b) = 1$. By (248) there exists a reduced residue system modulo a, say $n_1, \ldots, n_{\varphi(a)}$, and a reduced residue system modulo b, say $m_1, \ldots, m_{\varphi(b)}$, such that all the n's and m's are prime to ab. We would like to show that the set of all nm's is a reduced residue system modulo ab. In the language of (259) the nm's are the chickens and the congruence classes of integers prime to ab are the cages. There are $\varphi(a)\varphi(b)$ of the former and $\varphi(ab)$ of the latter. We want (d) of (259); it suffices to prove (a), (b), and (c).

If n and m are both prime to ab, then so is nm by (80). This says (a); every chicken is in a cage.

Proving (b) seems more troublesome. We must show that if $nm \equiv n'm' \pmod{ab}$, then $n = n'$ and $m = m'$. The congruence says that $ab|nm - n'm'$, but it is not clear how to proceed. Let us skip this and look at (c).

To prove (c) we must show that if k is any number prime to ab, then there exist n and m such that $nm \equiv k \pmod{ab}$. Let us examine the general problem of finding x such that $x \equiv k \pmod{ab}$. This congruence implies that $x \equiv k \pmod a$ and $x \equiv k \pmod b$. [This is (102e).] Does the converse hold? Let us suppose $x \equiv k \pmod a$ and $x \equiv k \pmod b$. Then $a|x - k$ and $b|x - k$. We would like to conclude that $ab|x - k$. Let $x - k = ac$. Then $b|ac$. But $(a,b) = 1$, so $b|c$ by (249). If $c = bd$, then $x - k = abd$. We see $x \equiv k \pmod{ab}$.

(261) **Proposition**. Suppose $(a,b) = 1$. Then $x \equiv k \pmod{ab}$ if and only if $x \equiv k \pmod a$ and $x \equiv k \pmod b$.

(262) **Exercise***. Suppose $(a, b) = 1$. Give a proof that if a and b divide h, then so does ab, based on the prime factorization of h.

(263) Even though our progress in proving that $\varphi(ab) = \varphi(a)\varphi(b)$ is slow, we are coming up with a number of interesting results about congruences. Let us illustrate (261) with $a = 3$ and $b = 4$. The integers $1, 2, \ldots, 12$ are a complete residue system modulo 12; let us examine how they fall into the congruence classes modulo 3 and 4. We will label the latter by the complete residue systems $1, 2, 3$ and $1, 2, 3, 4$.

12	3	4
1	1	1
2	2	2
3	3	3
4	1	4
5	2	1
6	3	2
7	1	3
8	2	4
9	3	1
10	1	2
11	2	3
12	3	4

We notice that in the 3 and 4 columns we get all possible ordered pairs (n, m) such that $1 \leqslant n \leqslant 3$ and $1 \leqslant m \leqslant 4$. We could have predicted this from (261). There are $12 = 3 \cdot 4$ possible ordered pairs (the cages) and 12 congruence classes modulo 12 (the chickens). By (261) if two numbers are congruent modulo both 3 and 4, i.e., if they produce the same ordered pair, then they are congruent modulo 12. Thus no two chickens are in the same cage. We see (d) and (b) of (259) are satisfied. Since every number produces *some* ordered pair, (a) is clear. Thus (c) follows and there is a one-to-one correspondence between chickens and cages.

(264) *Warning.* In (259) we assumed without saying so that no chicken can be in two different cages. This is reasonable for chickens and cages but may not be in some other application. Before (259) is used it should be checked out. The application above was justified, since each congruence class modulo 12 (chicken) gives rise to a unique pair of congruence classes modulo 3 and 4 (cage).

(265) If we could generalize the argument of (263) to an arbitrary pair of relatively prime integers a and b we could conclude that the simultaneous congruences

$$x \equiv n \pmod a$$
$$x \equiv m \pmod b$$

always had a solution. Let us suppose n runs through a complete residue system modulo a, m through a complete residue system modulo b, and k through a complete residue system modulo ab. We will match up the k's with the ordered pairs (n, m) by the rule $k \longleftrightarrow (n, m)$ if and only if $k \equiv n \pmod{a}$ and $k \equiv m \pmod{b}$. There are ab of the k's and the same number of ordered pairs. Clearly each k is associated with a unique ordered pair. Different k's go with different ordered pairs by (261). We conclude that all the ordered pairs must be used up.

(266) **Proposition.** Suppose $(a, b) = 1$. Let n and m be any two integers. Then the system

$$x \equiv n \pmod{a}$$
$$x \equiv m \pmod{b}$$

has exactly one solution in any complete residue system modulo ab. Any two solutions are congruent modulo ab.

(267) **Exercise.** Solve the system

$$x \equiv 2 \pmod{6}$$
$$x \equiv 3 \pmod{5}.$$

(268) **Exercise.** Solve the system

$$x \equiv 2 \pmod{3}$$
$$x \equiv 1 \pmod{4}$$
$$x \equiv 3 \pmod{5}.$$

27 THE FUNCTION φ PROVED MULTIPLICATIVE

(269) The matching up of a complete residue system modulo ab with ordered pairs from complete residue systems modulo a and b went so well above that we are tempted to try the same thing for *reduced* residue systems. This time let us suppose that k, n, and m run through reduced residue systems modulo ab, a, and b, respectively. There are $\varphi(ab)$ of the k's and $\varphi(a)\varphi(b)$ ordered pairs (n, m). If we can apply (259) to conclude part (d), that the number of chickens equals the number of cages, we will have our long-sought-after proof that φ is multiplicative. To do this we must establish parts (a), (b), and (c) of (259).

As before we match k with (n, m) in case $k \equiv n \pmod{a}$ and $k \equiv m \pmod{b}$. Since $(k, ab) = 1$, we have $(k, a) = (k, b) = 1$ also. Thus each k matches with one and only one ordered pair. The matching we are using here is exactly the same as we used in (265), so we already know that no two k's match to the same ordered pair. We see (a) and (b) of (259) hold. (As before the k's are the chickens and the (n, m)'s are the cages. Clearly no chicken is in more than one cage.)

It remains to show (c) of (259); namely, that each (n, m) is matched with some k. Let n and m be given. By (266) there exists x such that $x \equiv n \pmod{a}$ and $x \equiv m \pmod{b}$. Since $(x, a) = (n, a) = 1$, and $(x, b) = (m, b) = 1$, we see $(x, ab) = 1$ by (80). Thus $x \equiv k \pmod{ab}$ for some k.

We finally have arrived at an independent proof that φ is multiplicative. Although this follows easily from the formula we derived for φ, and although the proof we found is not at all of the form we originally expected, the search for it was nevertheless rewarding.

(270) **Exercise.** Consider $a = 5$ and $b = 6$. Let the n's be $1, -1, 2, 8$; let the m's be $5, 7$; and let the k's be $1, 7, -1, -7, 11, -11, 13, -13$. Match each k with an (n,m) as in (269).

(271) **Exercise*.** Find reduced residue systems for 4 and 5, choosing the least positive integers relatively prime to 20. Call the systems the n's and m's respectively. Show that the nm's do *not* form a reduced residue system modulo 20. Show that if the n's are taken to be 1 and 11 and the m's are as before, then the new set of nm's is a reduced residue system modulo 20.

(272) The last exercise shows that, promising as it seemed, our original idea for proving $\varphi(ab) = \varphi(a)\varphi(b)$ (which would have ended with the line "thus the nm's are a reduced residue system modulo ab") still has some bugs to be worked out. No wonder we got bogged down in (260). Evidently it doesn't suffice merely that the n's be prime to b and the m's to a. Of course (248) tells us that we have complete freedom in choosing how we want the n's to act modulo b; likewise for the m's modulo a. Perhaps we should exercise this freedom fully. We can specify the n's to be anything we want modulo b. Let us take them all to be congruent to 1 modulo b; perhaps this will simplify things. Let us take the m's all congruent to 1 modulo a.

Now let us try to show the nm's are incongruent modulo ab. Suppose $nm \equiv n'm' \pmod{ab}$. Then $nm \equiv n'm' \pmod{a}$. But $m \equiv m' \equiv 1 \pmod{a}$, so we have $n \equiv n' \pmod{a}$. Thus $n = n'$. Likewise $m = m'$. Since the nm's are all incongruent modulo ab we can conclude that $\varphi(ab) \geqslant \varphi(a)\varphi(b)$.

It remains to show that the nm's hit *all* the residue classes modulo ab of numbers prime to ab. Suppose $(k,ab) = 1$. We want n and m such that $nm \equiv k \pmod{ab}$. By (261) it suffices that $nm \equiv k \pmod{a}$ and $nm \equiv k \pmod{b}$. By our choice of the n's and m's this is the same as $n \equiv k \pmod{a}$ and $m \equiv k \pmod{b}$. Of course such an n exists in any reduced residue system modulo a $((k,ab) = 1$ implies $(k,a) = 1)$; similarly for m. We have still another proof that φ is multiplicative.

(273) **Exercise.** Suppose $(a,b) = 1$, $(i,a) = 1$, and $(j,b) = 1$. Suppose the n's form a reduced residue system modulo a, all of whose members are congruent to j modulo b. Suppose the m's form a reduced residue system modulo b, all of whose members are congruent to i modulo a. Show the nm's are a reduced residue system modulo ab.

(274) The proof in (272) has the defect that we do not know *explicitly* what the n's and m's are. (Proposition (266) has a similar defect; it tells us x exists but it doesn't tell what it is.) Let us examine the n's and m's more closely. The m's are chosen so as to run through a reduced residue system modulo b while

simultaneously all are congruent to 1 modulo a. Thus each m equals $1 + xa$, for some integer x. In the same way each n equals $1 + yb$ for some y. Then

$$nm = 1 + xa + yb + xyab$$
$$\equiv 1 + xa + yb \pmod{ab}.$$

From this point of view we should not be surprised that the nm's run through a reduced residue system modulo ab. Way back in (57) we proved that if $(a,b) = 1$ then the form $ax + by$ represents *all* integers; of course the same holds for $1 + ax + by$. Since $ax + by \equiv by \pmod{a}$ and $ax + by \equiv ax \pmod{b}$, in order that $ax + by$ run through a reduced residue system modulo ab it suffices that ax and by run through reduced residue systems modulo b and a respectively. [This follows from (266).] By (251) it suffices that x and y do. Still yet another proof that φ is multiplicative.

(275) **Exercise.** Suppose $(a,b) = 1$. Suppose n runs through a reduced residue system modulo a and m runs through a reduced residue system modulo b. Show that $am + bn$ runs through a reduced residue system modulo ab, assuming no results past (235).

(276) **Exercise.** Prove (275) with "reduced" replaced by "complete".

(277) **Exercise.** Let $S = \{1,3\}$, $T = \{1,2,3,4\}$, $U = \{1,2,3,4,5\}$. Confirm that the set of elements $4t + 5s$, where s is in S and t is in T, is a reduced residue system modulo 20. Confirm that the set of elements $4u + 5t$, where t is in T and u is in U, is a complete residue system modulo 20.

(278) **Exercise.** Suppose that as n runs through a reduced residue system modulo a and m runs through a reduced residue system modulo b, then nm runs through a reduced residue system modulo ab. Show $(a,b) = 1$.

(279) **Exercise.** Prove (278) with "reduced" replaced by "complete".

(280) **Exercise.** Suppose $(a,b) = 1$. Suppose that S is a complete residue system modulo a. Suppose that for each n in S the set T_n is a complete residue system modulo b. Show that the set of all numbers of the form $am + bn$, where n is in S and m is in T_n, is a complete residue system modulo ab.

(281) **Exercise.** Suppose $(a,b) = 1$. For any n let T_n be the set of integers m such that

$$-bn/a < m \leqslant b - bn/a.$$

Prove that the set of all numbers of the form $am + bn$ where $1 \leqslant n \leqslant a$ and m is in T_n, is exactly the set of positive integers not exceeding ab. (For example, if $a = 4$ and $b = 5$, then $T_3 = \{-3, -2, -1, 0, 1\}$.)

chapter 4

Solving Congruences

An important part of algebra concerns finding solutions to equations and systems of equations, usually polynomials. In what follows some of the corresponding theory for congruences is developed.

28 THE CHINESE REMAINDER THEOREM

(282) The discussion in (274) shows how to break down the problem of finding k in the simultaneous congruences

$$k \equiv n \pmod{a}$$
$$k \equiv m \pmod{b}.$$

It suffices to solve $xa \equiv m \pmod{b}$ and $yb \equiv n \pmod{a}$. Then $k = ax + by$ works. For example, suppose the system

$$k \equiv 2 \pmod{6}$$
$$k \equiv 3 \pmod{5}$$

is given. Let $a = 6$, $b = 5$, $n = 2$, and $m = 3$. First we solve $6x \equiv 3 \pmod{5}$. A solution is $x = 3$. Next we solve $5y \equiv 2 \pmod{6}$. A solution is $y = 4$. Then $k = xa + yb = 3 \cdot 6 + 4 \cdot 5 = 38$ is a solution to the system of congruences. By (266) any number congruent to 38 modulo 30 is also a solution; 8, for example.

(283) **Exercise.** Find a solution k to each of the following systems of congruences, subject to the conditions given.
   (a) $k \equiv 2 \pmod{6}$
   $\quad k \equiv 3 \pmod{5}$, $-30 < k \leqslant 0$
   (b) $k \equiv 3 \pmod{4}$
   $\quad k \equiv 5 \pmod{7}$, $0 \leqslant k \leqslant 27$
   (c) $k \equiv 1 \pmod{100}$
   $\quad k \equiv 1 \pmod{101}$, $-10100 \leqslant k < 0$
   (d) $k \equiv 1 \pmod{4}$
   $\quad k \equiv 3 \pmod{5}$
   $\quad k \equiv 2 \pmod{3}$, $0 < k \leqslant 60$.

(284)   The question arises whether Proposition (266) can be extended to more than two simultaneous congruences.   Let us try three, just to get started. Consider the system

$$x \equiv n \pmod{a}$$
$$x \equiv m \pmod{b}$$
$$x \equiv l \pmod{c}.$$

If $(a,b) = 1$ we know by (266) that the first two of these congruences have a simultaneous solution, say x_0, and that all solutions (of these two) are just the numbers congruent to x_0 modulo ab. Thus our original system is equivalent to

$$x \equiv x_0 \pmod{ab}$$
$$x \equiv l \quad \pmod{c}.$$

For this to be solvable it suffices that $(ab,c) = 1$, again by (266). By (80) this is the same as $(a,c) = 1$ and $(b,c) = 1$.

(285)   **Definition.** We say the integers $m_1, m_2, \ldots, m_t$ are *relatively prime in pairs* in case $(m_i, m_j) = 1$ whenever $i \neq j$.

(286)   **The Chinese Remainder Theorem.** Suppose the positive integers m_1, $m_2, \ldots, m_t$ are relatively prime in pairs. Let $b_1, \ldots, b_t$ be any integers. Then the system of congruences

$$x \equiv b_1 \pmod{m_1}$$
$$x \equiv b_2 \pmod{m_2}$$
$$\cdots$$
$$x \equiv b_t \pmod{m_t}$$

has a solution x_0.   Moreover, x is a solution if and only if $x \equiv x_0 \pmod{m_1 m_2 \ldots m_t}$.

Proof. The proof will be by induction on the number t of congruences [see (169) (a)]. For $t = 1$ there is nothing to prove. The case $t = 2$ is (266). Let us assume the theorem is true for $t - 1$. Then there exists a number y_0 such that the $t - 1$ congruences

$$x \equiv b_1 \pmod{m_1}$$
$$\cdots$$
$$x \equiv b_{t-1} \pmod{m_{t-1}}$$

are equivalent to the single congruence

$$x \equiv y_0 \pmod{m_1 m_2 \ldots m_{t-1}}.$$

Thus our original system of t congruences is equivalent to

$$x \equiv y_0 \pmod{m_1 m_2 \ldots m_{t-1}}$$
$$x \equiv b_t \pmod{m_t}.$$

We leave it to the reader to prove that $(m_1 m_2 \ldots m_{t-1}, m_t) = 1$. [See (287).] Thus (266) can be applied to conclude that this pair of congruences (and so the original system) has a solution x_0, and that x is a solution if and only if $x \equiv x_0$ (mod $m_1 m_2 \ldots m_t$).

(287) **Exercise*.** Suppose $(m_i, m) = 1$ for $i = 1, 2, \ldots, k$. Prove
$$(m_1 m_2 \ldots m_k, m) = 1.$$

(288) **Exercise.** Solve the system

$$x \equiv 1 \quad (\text{mod } 2)$$
$$x \equiv 2 \quad (\text{mod } 3)$$
$$x \equiv -1 \ (\text{mod } 5)$$
$$x \equiv 3 \quad (\text{mod } 7), 1 \leqslant x \leqslant 210.$$

[*Hint:* The method of (282) may be used to combine the first two congruences into one, with modulus 6. Thus the four congruences are reduced to three. This can be repeated until a single congruence remains. Compare with our proof of the Chinese Remainder Theorem.]

(289) **Exercise.** Solve

$$x \equiv 3 \quad (\text{mod } 8)$$
$$x \equiv 2 \quad (\text{mod } 7)$$
$$x \equiv 1 \quad (\text{mod } 5), 1 \leqslant x \leqslant 280.$$

(290) **Exercise.** Solve

$$x \equiv -2 \quad (\text{mod } 9)$$
$$x \equiv 0 \quad \quad (\text{mod } 5)$$
$$x \equiv 3 \quad \quad (\text{mod } 4), -90 < x \leqslant 90.$$

(291) **Exercise.** Solve

$$x \equiv 1 \quad (\text{mod } 3)$$
$$x \equiv 2 \quad (\text{mod } 4)$$
$$x \equiv 3 \quad (\text{mod } 10), 1 \leqslant x \leqslant 120.$$

29 AN IMPROVED VERSION OF THE CHINESE REMAINDER THEOREM

(292) In (282) we found an explicit formula for solving the simultaneous congruences

$$\xi \equiv n \quad (\text{mod } a)$$
$$\xi \equiv m \quad (\text{mod } b),$$

namely $\xi = xa + yb$, where $xa \equiv m$ (mod b) and $yb \equiv n$ (mod a). We may wonder if there is a similar formula that would enable us to solve three or more congruences all at once, rather than stepwise as suggested in Exercise (288).

Let us see what we can do with the three congruences

(*)
$$\xi \equiv n \pmod{a}$$
$$\xi \equiv m \pmod{b}$$
$$\xi \equiv l \pmod{c}.$$

Analogy with the case of two congruences suggests trying a solution of the form $\xi = xa + yb + zc$. Then x, y, and z must somehow be chosen such that

$$yb + zc \equiv n \pmod{a},$$
$$xa + zc \equiv m \pmod{b},$$

and

$$xa + yb \equiv l \pmod{c}.$$

This looks pretty complicated; perhaps we are on the wrong track.

The reader may care to go back to (274) to see where we came up with the form $xa + yb$ in the first place. There we were searching for a form that ran through a complete residue system modulo ab, and settled on nm, where n and m ran through complete residue systems modulo a and b, respectively, while $n \equiv 1 \pmod{b}$ and $m \equiv 1 \pmod{a}$ for all n and m. Then $nm = (yb + 1)(xa + 1) \equiv xa + yb + 1 \pmod{ab}$ for some x and y; and we dropped the superfluous 1.

In the same way we may choose n', m', and l' such that $n' \equiv n \pmod{a}$, $n' \equiv 1 \pmod{b}$, and $n' \equiv 1 \pmod{c}$, with similar congruences holding for m' and l'. By (261), $n' \equiv 1 \pmod{bc}$. (We are assuming, of course, that a, b, and c are relatively prime in pairs.) Thus $n' = xbc + 1$ for some x; likewise $m' = yac + 1$ and $l' = zab + 1$. We compute $n'm'l'$, remembering that we are only interested in solutions modulo abc: $n'm'l' \equiv xbc + yac + zab + 1 \pmod{abc}$.

Let us examine under what circumstances $\xi = xbc + yac + zab$ will be a solution of (*). We need

$$\xi \equiv xbc \equiv n \pmod{a}$$
$$\xi \equiv yac \equiv m \pmod{b}$$
$$\xi \equiv zab \equiv l \pmod{c}.$$

Here x, y, and z are the unknowns. Each of these congruences has a solution according to (251).

(293) **Example.** Consider the three congruences

$$\xi \equiv 1 \pmod{3}$$
$$\xi \equiv 2 \pmod{4}$$
$$\xi \equiv -1 \pmod{7}.$$

Here $a = 3, b = 4, c = 7$. First we solve the congruences

$$x \cdot 4 \cdot 7 \equiv 1 \pmod{3}$$
$$y \cdot 3 \cdot 7 \equiv 2 \pmod{4}$$
$$z \cdot 3 \cdot 4 \equiv -1 \pmod{7}.$$

These can be simplified to

$$x \equiv 1 \pmod 3$$
$$y \equiv 2 \pmod 4$$
$$5z \equiv -1 \pmod 7.$$

(Since $4 \cdot 7 = 28 \equiv 1 \pmod 3$, etc.) By inspection we get the solutions $x = 1$, $y = 2, z = 4$. Then

$$\xi = 1 \cdot 4 \cdot 7 + 2 \cdot 3 \cdot 7 + 4 \cdot 3 \cdot 4 = 118$$

is a solution to the original system. By the Chinese Remainder Theorem so is anything congruent to 118 modulo $abc = 84$; a smaller solution is 34. This is easily checked to be correct.

(294) **Exercise.** Use the above method to solve

$$\xi \equiv 1 \pmod 2$$
$$\xi \equiv -2 \pmod 5$$
$$\xi \equiv 3 \pmod 7, 1 \leqslant \xi \leqslant 70.$$

(295) We will see later that the solution of more complicated congruences sometimes involves applying the Chinese Remainder Theorem to systems such as

$$\xi \equiv 2 \qquad\qquad \pmod 3$$
$$\xi \equiv 2 \text{ or } 3 \qquad \pmod 4$$
$$\xi \equiv 0, 1, \text{ or } -1 \pmod 5,$$

where more than one congruence class may be allowed for individual moduli. Here $a = 3$, $b = 4$, and $c = 5$; as before we seek a solution of the form $\xi = xbc + yac + zab$. We need

$$x \cdot 4 \cdot 5 \equiv 2 \qquad\qquad \pmod 3$$
$$y \cdot 3 \cdot 5 \equiv 2 \text{ or } 3 \qquad \pmod 4$$
$$z \cdot 3 \cdot 4 \equiv 0, 1, \text{ or } -1 \pmod 5.$$

The first congruence has a solution $x = 1$. We could solve the congruences

$$15y \equiv 2 \pmod 4$$
and $$15y \equiv 3 \pmod 4$$

individually, but there is a way to combine the work. We consider the single congruence

$$15y' \equiv 1 \pmod 4.$$

If y' is a solution of this, then $2y' \equiv 2 \pmod 4$ and $3y' \equiv 3 \pmod 4$. Thus we need solve only one congruence instead of two. Here $y' = -1$ generates the solutions $y = -2$ and $y = -3$ to $15y \equiv 2$ or $3 \pmod 4$.

In the same way we consider

$$12z' \equiv 1 \pmod 5.$$

A solution is $z' = 3$. Thus $z = 0, 3$, and -3 are solutions to

$$12z \equiv 0, 1, \text{ or } -1 \pmod 5.$$

As in (292) we find $\xi = 20x + 15y + 12z$ is a solution to the original system. We calculate

$$\xi = 20(1) + 15(-2) + 12(0) = -10$$
$$\xi = 20(1) + 15(-3) + 12(0) = -25$$
$$\xi = 20(1) + 15(-2) + 12(3) = 26$$
$$\xi = 20(1) + 15(-3) + 12(3) = 11$$
$$\xi = 20(1) + 15(-2) + 12(-3) = -46$$
$$\xi = 20(1) + 15(-3) + 12(-3) = -61.$$

If we desire solutions in the range $0 \leqslant \xi < 60$ these reduce to $\xi = 50, 35, 26, 11, 14, 59$.

(296) **Exercise.**

   (a)   Solve $5x \equiv 2, 3,$ or $-1 \pmod 7, 0 \leqslant x < 7$.
   (b)   Solve $100\,x \equiv 3$ or $14 \pmod{11}, 1 \leqslant x \leqslant 11$.
   (c)   Solve $11x \equiv 10, 19,$ or $-2 \pmod{25}, |x| \leqslant 12$.

(297) **Exercise.** Solve

$$\xi \equiv 1 \text{ or } 2 \pmod{10}$$
$$\xi \equiv 2 \text{ or } 3 \pmod 7$$
$$\xi \equiv 3 \text{ or } 4 \pmod 3, 0 \leqslant \xi < 210.$$

(298) **Exercise.** Solve

$$\xi \equiv 1 \text{ or } 2 \pmod 2$$
$$\xi \equiv 3 \pmod 5$$
$$\xi \equiv 5, 6, \text{ or } 7 \pmod 9, 0 < \xi \leqslant 90.$$

(299) **The Chinese Remainder Theorem** (improved version). Suppose the positive integers $m_1, m_2, \ldots, m_t$ are relatively prime in pairs. Let $b_1, \ldots, b_t$ be any integers. Then the system of congruences

$$\xi \equiv b_1 \pmod{m_1}$$
$$\xi \equiv b_2 \pmod{m_2}$$
$$\cdots$$
$$\xi \equiv b_t \pmod{m_t}$$

has a solution $\xi_0 = \sum_{i=1}^{t} (M/m_i)b_i x_i'$, where $M = m_1 m_2 \ \ldots \ m_t$ and $(M/m_i)x_i' \equiv 1 \pmod{m_i}$, $i = 1, 2, \ldots, t$. Moreover, ξ is a solution if and only if $\xi \equiv \xi_0 \pmod M$.

   Proof. We note that $(M/m_i, m_i) = 1$ by (287); thus (251) guarantees the existence of x_i' such that $(M/m_i) x_i' \equiv 1 \pmod{m_i}$ for each i. It remains to show ξ_0 satisfies our system of congruences. Consider the ith one. Since $m_i | M/m_j$ for

$j \neq i$, we have

$$\xi_0 = \sum_{j=1}^{t} (M/m_j) b_j x_j{}' \equiv (M/m_i) b_i x_i{}' \equiv b_i \ (\text{mod } m_i).$$

This works for $i = 1, 2, \ldots, t$; ξ_0 is a solution.

(300) **Example.** Let us solve

$$\begin{aligned}
\xi &\equiv 0 \ (\text{mod } 2) \\
\xi &\equiv 3 \ (\text{mod } 5) \\
\xi &\equiv -1 \ (\text{mod } 3) \\
\xi &\equiv 2 \ (\text{mod } 7).
\end{aligned}$$

Let $m_1 = 2, m_2 = 5, m_3 = 3, m_4 = 7$. $M = 2 \cdot 5 \cdot 3 \cdot 7 = 210$. We solve

$$\begin{aligned}
5 \cdot 3 \cdot 7 x_1' &= 105 x_1' \equiv x_1' \equiv 1 \ (\text{mod } 2) \\
2 \cdot 3 \cdot 7 x_2' &= 42 x_2' \equiv 2 x_2' \equiv 1 \ (\text{mod } 5) \\
2 \cdot 5 \cdot 7 x_3' &= 70 x_3' \equiv x_3' \equiv 1 \ (\text{mod } 3) \\
2 \cdot 5 \cdot 3 x_4' &= 30 x_4' \equiv 2 x_4' \equiv 1 \ (\text{mod } 7).
\end{aligned}$$

Solutions are $x_1' = 1, x_2' = 3, x_3' = 1, x_4' = 4$. Then $\xi_0 = 105 \cdot 1 \cdot 0 + 42 \cdot 3 \cdot 3 + 70 \cdot 1(-1) + 30 \cdot 4 \cdot 2 = 548 \equiv 128 \ (\text{mod } 210)$.

(301) **Exercise.*** Solve

$$\begin{aligned}
\xi &\equiv 2 \ (\text{mod } 3) \\
\xi &\equiv 3 \ (\text{mod } 5) \\
\xi &\equiv 2 \ (\text{mod } 7)
\end{aligned}$$

   by the method of (299).

(302) *Note.* The Chinese writer Sun-Tse solved (301) in the same way you just did about the first century A.D.

(303) **Exercise.** Solve

$$\begin{aligned}
\alpha &\equiv 3 \ (\text{mod } 9) \\
\alpha &\equiv 1 \ (\text{mod } 2) \\
\alpha &\equiv 1 \ (\text{mod } 5) \\
\alpha &\equiv 0 \ (\text{mod } 7), 0 \leqslant \alpha < 630.
\end{aligned}$$

(304) **Exercise.** Solve

$$\begin{aligned}
\theta &\equiv 2 \ (\text{mod } 15) \\
\theta &\equiv 2 \ (\text{mod } 4) \\
\theta &\equiv 2 \ (\text{mod } 7), -420 < \theta \leqslant 0.
\end{aligned}$$

(305) **Exercise.** Solve

$$\begin{aligned}
k &\equiv 1 \ (\text{mod } 2) \\
k &\equiv 5 \ (\text{mod } 3) \\
k &\equiv 2 \ (\text{mod } 5) \\
k &\equiv 0 \ \text{or} \ 1 \ (\text{mod } 7), 0 \leqslant k < 210.
\end{aligned}$$

(306) **Exercise**. Professor Crittenden buys a new car every three years; he bought his first in 1961. He gets a sabbatical leave every seven years, starting in 1972. When will he first get both during a leap year?

(307) **Exercise**. Senator McKinley was first elected in 1962. His reelection is assured unless the campaign coincides with an attack of the seven-year itch such as hit him in 1965. When must he worry first?

30 THE CONGRUENCE $ax \equiv 1 \pmod b$

(308) The justification for calling (299) an "improved version" of the Chinese Remainder Theorem is that it does not merely assert the *existence* of a solution to the system of congruences, but goes on to tell explicitly how to find it. This explicitness extends only to a point, however, for we still find ourselves solving congruences of the form $ax \equiv 1 \pmod b$ by trial and error in the course of applying (299). This can be tedious if b is large, as it must be if there are many congruences in our system.

That the congruence $ax \equiv 1 \pmod b$ has a solution at all is guaranteed by (251), under the hypothesis that $(a,b) = 1$. We naturally look back to the proof of (251) [which is in (247)] to see if it contains any method of actually constructing a solution x. It does not.

(309) *Note*. A proof such as in (247) is called an "existence proof," as opposed to a "constructive proof." Often a theorem may have proofs of both types. Consider, for example

THEOREM. If b is any real number there exists a real number x such that $x^2 + bx - 1 = 0$.

Constructive proof. Let $x = -\frac{1}{2}(b + \sqrt{b^2 + 4})$. Substitution shows x satisfies the equation.

Existence proof. Let $f(x) = x^2 + bx - 1$. Then $f(0) = -1$, while $f(x) > 0$ for x sufficiently large. Since f is a continuous function there exists x such that $f(x) = 0$.

The proof in (247) is really not so "existence" as it could be, since it assures a solution to $ax \equiv 1 \pmod b$ in any complete residue system modulo b. Thus examining ax for $x = 0, 1, \ldots, b - 1$ would lead us to a solution in a finite number of steps.

(310) We return to the problem of solving $ax \equiv 1 \pmod b$, where $(a,b) = 1$. We want an x such that $ax - 1 = yb$, for some integer y. Writing this as $xa + y(-b) = 1$ should ring a bell. We solved the problem of finding x and y such that $xa +$

$yb = (a,b)$ back in (49); moreover, we saw how to determine x and y *explicitly* by solving backwards the equations generated in using the Euclidean Algorithm to find (a,b).

(311) **Examples.** Consider the congruence $15x \equiv 1 \pmod{22}$. We apply the Euclidean Algorithm to 15 and 22, as follows:

$$22 = 1 \cdot 15 + 7$$
$$15 = 2 \cdot 7 + 1$$
$$7 = 7 \cdot 1 + 0.$$

Thus $1 = 15 - 2 \cdot 7 = 15 - 2(22 - 15) = 3 \cdot 15 - 2 \cdot 22$. We see $3 \cdot 15 \equiv 1 \pmod{22}$. Thus $x = 3$ is a solution.

Suppose we want to solve $321x \equiv 7 \pmod{1000}$. First we consider $321x' \equiv 1 \pmod{1000}$. We have

$$1000 = 3 \cdot 321 + 37$$
$$321 = 8 \cdot 37 + 25$$
$$37 = 1 \cdot 25 + 12$$
$$25 = 2 \cdot 12 + 1.$$

Solving these equations backwards gives us $1 = 81 \cdot 321 - 26 \cdot 1000$. Thus $321 \cdot 81 \equiv 1 \pmod{1000}$; $x' = 81$. Then $321(7 \cdot 81) \equiv 7 \pmod{1000}$; $x = 567$.

When used on two simultaneous congruences, this method kills two birds with one stone. Consider, for example,

$$\xi \equiv 2 \pmod{321}$$
$$\xi \equiv 3 \pmod{1000}.$$

We have seen that $81 \cdot 321 - 26 \cdot 1000 = 1$. Thus $-26 \cdot 1000 \equiv 1 \pmod{321}$ and $81 \cdot 321 \equiv 1 \pmod{1000}$. Then

$$\xi = 2 \cdot (-26) \cdot 1000 + 3 \cdot 81 \cdot 321 = 26003$$

is a solution to the system.

(312) **Exercise.** Find a solution x to each of the following congruences satisfying $0 \leqslant x < m$, where m is the modulus.
   (a) $15x \equiv 1 \pmod{41}$
   (b) $300x \equiv 1 \pmod{401}$
   (c) $91x \equiv 25$ or $35 \pmod{121}$.

(313) **Exercise.** Solve

$$\xi \equiv 4 \pmod{303}$$
$$\xi \equiv -3 \pmod{350}, 0 \leqslant \xi < 106050.$$

(314) **Exercise**. Solve

$$x \equiv 2 \,(\text{mod } 7)$$
$$x \equiv 3 \,(\text{mod } 5)$$
$$x \equiv 1 \,(\text{mod } 4)$$
$$x \equiv -3 \,(\text{mod } 11), \, 0 \leqslant x < 1540.$$

31 POLYNOMIAL CONGRUENCES

(315) According to (251) the congruence $ax \equiv h \,(\text{mod } b)$ has exactly one solution in any complete residue system modulo b, assuming $(a,b) = 1$. Moreover, if $y \equiv x \,(\text{mod } b)$, then x and y are either both or neither solutions of the congruence. The latter statement is true whether $(a,b) = 1$ or not, though if $(a,b) > 1$ the congruence may have more than one solution in a complete residue system modulo b. (For example $2x \equiv 4 \,(\text{mod } 6)$ has 2 and 5 as solutions.) The proof of the following theorem is an easy consequence of (102), and is left to the reader.

(316) **THEOREM.** If f is a polynomial with integral coefficients and $x \equiv y \,(\text{mod } m)$, then $f(x) \equiv f(y) \,(\text{mod } m)$. In particular $f(x) \equiv 0 \,(\text{mod } m)$ if and only if $f(y) \equiv 0 \,(\text{mod } m)$.

(317) **True-False.** Let x, y and m be positive integers, with $x \equiv y \,(\text{mod } m)$.
  (a) Suppose f is a polynomial (not necessarily with integral coefficients) such that $f(x)$ and $f(y)$ are integers. Then $f(x) \equiv f(y) \,(\text{mod } m)$.
  (b) $x! \equiv y! \,(\text{mod } m)$.
  (c) $2^x \equiv 2^y \,(\text{mod } m)$.
  (d) $x^x \equiv y^y \,(\text{mod } m)$.

(318) **Exercise.** Prove a necessary and sufficient condition that $2^a \equiv 2^b$ $(\text{mod } 7)$.

(319) In order to save words let us make the convention that from now on by *polynomial* we mean *polynomial with integral coefficients* unless the contrary is expressly stated; integral polynomials are the most natural to study from the number-theoretic point of view.

Theorem (316) suggests another word-saving convention. It says that if f is a polynomial then the solutions of the congruence $f(x) \equiv 0 \,(\text{mod } m)$ consist of entire congruence classes modulo m. Thus if we know all the solutions in some complete residue system modulo m we can deduce the rest. This motivates the following special meaning of "solution."

(320) **Definition.** If f is an (integral) polynomial we mean by a *complete solution* of the congruence $f(x) \equiv 0 \,(\text{mod } m)$ the set of all solutions of the con-

gruence in any complete residue system modulo m. More generally, if f_1, f_2, ..., f_t are polynomials and $m_1, m_2, \ldots, m_t$ are positive integers, relatively prime in pairs, by a *complete solution* of the system

$$f_1(x) \equiv 0 (\text{mod } m_1)$$
$$f_2(x) \equiv 0 (\text{mod } m_2)$$
$$\cdots$$
$$f_t(x) \equiv 0 (\text{mod } m_t)$$

we mean the set of all solutions of the system in any complete residue system modulo $m_1 m_2 \ldots m_t$.

(321) **Examples.** A complete solution of $3x \equiv 2$ (mod 5) (which is equivalent to $3x - 2 \equiv 0$ (mod 5)) is $x = 4$, since it comprises all the solutions among 0, 1, 2, 3, 4; another complete solution is $x = -1$. A complete solution of $3x \equiv 0$ (mod 6) is $x = 0, 2, 4$; another is $x = -2, 0, 2$. A complete solution to $x^2 - x \equiv 0$ (mod 6) is $x = 0, 1, 3, 4$. A complete solution to

$$x^2 \equiv 0 (\text{mod } 3)$$
$$x^2 \equiv 0 (\text{mod } 4)$$

is $x = 0, 6$; it comprises all $x, 0 \leqslant x < 12$, which are solutions.

(322) **Exercise.** Give complete solutions consisting of even integers to each
of the following
(a) $17x \equiv 13 (\text{mod } 25)$
(b) $6y \equiv 3 (\text{mod } 15)$
(c) $z \equiv 2 (\text{mod } 11)$
$z \equiv 3 (\text{mod } 13)$.

32 SOLVING $ax \equiv c$ (mod b) IN GENERAL

(323) Even if $(a,b) > 1$ the Euclidean Algorithm method still enables us to calculate x and y such that $ax + by = (a,b)$. Then $ax \equiv (a,b)$ (mod b), and $a(kx) \equiv k(a,b)$ (mod b) for any integer k. This provides a solution to $ax \equiv c$ (mod b) whenever c is a multiple of (a, b).

Consider, for example, $6x \equiv 10$ (mod 14). We write

$$14 = 2 \cdot 6 + 2$$
$$6 = 3 \cdot 2 + 0.$$

Thus $(14,6) = 2 = 1 \cdot 14 - 2 \cdot 6$, and $-2 \cdot 6 \equiv 2$ (mod 14). Then $5(-2) \cdot 6 \equiv 10$ (mod 14). We see -10 is a solution to the original congruence; 4 is another.

Two questions arise: $1°$ What if $(a,b) \nmid c$? $2°$ Does this method give a *complete* solution?

The first question is easily answered. If $ax \equiv c \pmod{b}$ then $ax - c = kb$ for some integer k. Clearly $(a,b)|c$. We conclude that if $(a,b) \nmid c$ the congruence $ax \equiv c \pmod{b}$ has no solution.

Let us now consider the second question. When $(a,b) = 1$ Theorem (251) tells us any complete solution of $ax \equiv c \pmod{b}$ has exactly one element. Suppose $(a,b) > 1$ and $(a,b)|c$. If we could divide the congruence (including the modulus) through by (a,b), then we could apply (251). The reader should have no trouble proving the following.

(324) **Proposition.** Let x, y, k, and m be integers, m and k positive. Then $x \equiv y \pmod{m}$ if and only if $kx \equiv ky \pmod{km}$.

(325) By (324) the congruence $ax \equiv c \pmod{b}$ is equivalent to $(a/d)x \equiv c/d \pmod{b/d}$, where $d = (a,b)$. We know there exist x and y such that $ax + by = d$, so $(a/d)x + (b/d)y = 1$. This equation implies $(a/d, b/d) = 1$. [See (59).] Thus we can invoke (251) to conclude that the congruence $(a/d)x \equiv c/d \pmod{b/d}$ has exactly one solution in any complete residue system modulo b/d.

As an example, let us look again at $6x \equiv 10 \pmod{14}$. Here $(a,b) = (6,14) = 2$. Thus an equivalent congruence is $3x \equiv 5 \pmod{7}$. A solution of this is $x = 4$; so is x' if and only if $x' \equiv 4 \pmod{7}$. This means $x = 4$ is a complete solution of $3x \equiv 5 \pmod{7}$. On the other hand $x = 4$ is *not* a complete solution of $6x \equiv 10 \pmod{14}$. The catch is that the definition of complete solution depends on the modulus. A complete solution of $6x \equiv 10 \pmod{14}$ must contain all solutions in some complete residue system modulo 14, for example $0, 1, 2, \ldots, 13$. Among these integers both 4 and 11 are congruent to 4 modulo 7. Thus $x = 4, 11$ is a complete solution of $6x \equiv 10 \pmod{14}$.

(326) The preceding paragraph shows that in order to complete our analysis of $ax \equiv c \pmod{b}$ we must determine all solutions of $x \equiv x_0 \pmod{b/d}$ in some complete residue system modulo b, where

$$(a/d)x_0 \equiv c/d \pmod{b/d}.$$

We have $x \equiv x_0 \pmod{b/d}$ if and only if $x = x_0 + k(b/d)$ for some integer k. Furthermore, $x_0 + k(b/d) \equiv x_0 + k'(b/d) \pmod{b}$ if and only if $b|(k - k')b/d$, which says $d|k - k'$. Thus the integers

$$x_0, x_0 + b/d, x_0 + 2b/d, \ldots, x_0 + (d - 1)b/d$$

are all incongruent modulo b; and every integer congruent to x_0 modulo b is congruent to one of them. They form a complete solution of $ax \equiv c \pmod{b}$.

(327) **THEOREM.** Let $(a,b) = d$. Then the congruence

$$ax \equiv c \pmod{b}$$

has no solution if $d \nmid c$. If $d|c$, then the congruence has d solutions in any complete residue system modulo b; and if x_0 is any particular solution of the congruence, $x = x_0, x_0 + b/d, x_0 + 2b/d, \ldots, x_0 + (d-1)b/d$ is a complete solution.

(328)   The following routine may be used to solve $ax \equiv c \pmod b$. First the Euclidean Algorithm is used to determine (a,b). At this point we test whether $(a,b)|c$ or not; if it doesn't, there is no solution. If it does we solve our Euclidean Algorithm equations backwards to determine x and y such that $ax + by = (a,b)$. Then $ax \equiv (a,b) \pmod b$. Thus

$$a(xc/(a,b)) \equiv (a,b)c/(a,b) = c \pmod b.$$

Let $xc/(a,b) = x_0$. A complete solution can now be generated as in the theorem.

(329)   **Examples.** Consider $140x \equiv 133 \pmod{301}$. We have

$$\begin{aligned} 301 &= 2 \cdot 140 + 21 \\ 140 &= 6 \cdot 21 + 14 \\ 21 &= 1 \cdot 14 + 7 \\ 14 &= 2 \cdot 7 + 0. \end{aligned}$$

Thus $(140,301) = 7$. We note that $133/7 = 19$; solutions exist. Solving the equations backwards gives

$$7 = 7 \cdot 301 - 15 \cdot 140.$$

Thus $-15 \cdot 140 \equiv 7 \pmod{301}$. Then $19(-15) \cdot 140 \equiv 19 \cdot 7 = 133 \pmod{301}$; we take $x_0 = 19(-15) = -285$. Here $b/(a,b) = 301/7 = 43$, so a complete solution is

$$x = -285, -285 + 43, -285 + 2 \cdot 43, \ldots, -285 + 6 \cdot 43,$$

or

$$x = -285, -242, -199, -156, -113, -70, -27.$$

If we desired the smallest nonnegative complete solution it would be more convenient to find first the smallest nonnegative x_0' such that $x_0' \equiv x_0 \pmod{43}$ and start from there. Here $x_0' = 16$, and another complete solution is

$$x = 16, 59, 102, 145, 188, 231, 274.$$

As another example we consider the congruence $1485x \equiv 999 \pmod{2222}$. Here

$$\begin{aligned} 2222 &= 1 \cdot 1485 + 737 \\ 1485 &= 2 \cdot 737 + 11 \\ 737 &= 67 \cdot 11 + 0, \end{aligned}$$

so $(1485,2222) = 11$. But $11 \nmid 999$; there are no solutions.

(330)   **Exercise.** Find a complete solution to

$$246x \equiv 192 \pmod{558}.$$

(331) **Exercise.** Find a complete solution to

$$4100x \equiv 1800 \ (\text{mod } 5300).$$

(332) **Exercise.** Find a complete solution to

$$549x \equiv 345 \ (\text{mod } 909).$$

(333) **Exercise.** Find a complete solution to

$$616,921,415x \equiv 125164 \ (\text{mod } 356,192,826).$$

[*Hint:* see (104).]

(334) **Exercise.** Prove directly from the definition of the greatest common divisor that $(a/(a,b), b/(a,b)) = 1$.

33 REDUCTION TO PRIME POWER MODULI

(335) Theorem (327) provides a complete analysis of the congruence $ax \equiv c \ (\text{mod } b)$. This is the simplest case of the general problem of solving polynomial congruences of the form $f(x) \equiv 0 \ (\text{mod } m)$, namely, the case when $f(x)$ is the linear polynomial $ax - c$. No completely general method is known for solving $f(x) \equiv 0 \ (\text{mod } m)$, where f is an arbitrary polynomial, but some simplifications may be possible. Recall that (261) says that for $(m_1, m_2) = 1$, we have $i \equiv j$ (mod $m_1 m_2$) if and only if $i \equiv j$ (mod m_1) and $i \equiv j$ (mod m_2). Thus if we can write $m = m_1 m_2$, with m_1 and m_2 relatively prime, we can replace the congruence $f(x) \equiv 0 \ (\text{mod } m)$ by the pair of congruences

$$f(x) \equiv 0 \ (\text{mod } m_1)$$
$$f(x) \equiv 0 \ (\text{mod } m_2).$$

Solving these may be easier; at least the moduli are smaller. The solutions will consist of certain whole congruence classes modulo m_1 and m_2. The *simultaneous* solutions can then be computed by means of the Chinese Remainder Theorem.

(336) **Examples.** Consider $x^2 + 3x \equiv 0 \ (\text{mod } 12)$. We replace this by the congruences

$$x^2 + 3x \equiv 0 \ (\text{mod } 3)$$
$$x^2 + 3x \equiv 0 \ (\text{mod } 4).$$

By trial and error we find that $x = 0$ is a complete solution of the congruence modulo 3 and $x = 0, 1$ is a complete solution of the congruence modulo 4. Thus $x^2 + 3x \equiv 0 \ (\text{mod } 12)$ if and only if

$$x \equiv 0 \ (\text{mod } 3)$$
$$x \equiv 0 \text{ or } 1 \ (\text{mod } 4).$$

Applying the Chinese Remainder Theorem [as in (295)] we find $x = 0, 9$ is a complete solution modulo 12.

Now consider $x^2 + x + 1 \equiv 0 \pmod{12}$. Again we treat

$$x^2 + x + 1 \equiv 0 \pmod 3$$
$$x^2 + x + 1 \equiv 0 \pmod 4.$$

A complete solution of the congruence with modulus 3 is $x = 1$, but the congruence with modulus 4 has no solutions. We conclude that neither does the original congruence.

(337) **Exercise.** Find a complete solution to

$$x^2 - x \equiv 0 \pmod{35}.$$

(338) **Exercise.** Find a complete solution to

$$x^2 + 3x \equiv 0 \pmod{45}.$$

(339) **Exercise.** Find a complete solution to

$$x^2 - 1 \equiv 0 \pmod{24}.$$

(340) **Exercise.** Suppose $m = m_1 m_2$, with $(m_1, m_2) = 1$, and suppose f is a polynomial. Show that if the number of elements in complete solutions of $f(x) \equiv 0$ modulo m, m_1, and m_2 are k, k_1, and k_2 respectively, then $k = k_1 k_2$.

(341) **Exercise.** Compute the number of elements in complete solutions of $x^2 + 3x \equiv 0$ (mod 18, 3, and 6). Does this contradict (340)?

(342) Of course we are not limited to breaking m up into only two factors. If $m = m_1 m_2 \ldots m_t$, with the m_i relatively prime in pairs, the same technique of solving $f(x) \equiv 0 \pmod{m_i}$, $i = 1, 2, \ldots, t$, and then combining the results by means of the Chinese Remainder Theorem works. Since no prime can divide two different m_i, the furthest this can be carried is to break m into its prime power components.

(343) **THEOREM.** Suppose f is a polynomial and $m = p_1^{\alpha_1} \ldots p_t^{\alpha_t}$, where the p_i are distinct primes. Then x is a solution of $f(x) \equiv 0 \pmod m$ if and only if x is a solution of $f(x) \equiv 0 \pmod{p_i^{\alpha_i}}$ for $i = 1, 2, \ldots, t$. If a complete solution of $f(x) \equiv 0 \pmod{p_i^{\alpha_i}}$ has k_i elements, $i = 1, 2, \ldots, t$, then a complete solution of $f(x) \equiv 0 \pmod m$ has $k_1 k_2 \ldots k_t$ elements.

(344) *Note.* The proof of the above theorem is a straightforward generalization of the argument given in (335) for the case $t = 2$; we leave it to the reader. He will need a souped-up version of (261). This is given in Exercise (345), which should be proved first. The possibility that some of the k_i may be 0 is not excluded in the theorem.

(345) **Exercise***. Suppose $m = m_1 m_2 \ldots m_t$ and $(m_i, m_j) = 1$ when $i \neq j$. Prove in two ways that $x \equiv x' \pmod{m}$ if and only if $x \equiv x' \pmod{m_i}$, $i = 1, 2, \ldots, t$: $1°$ by induction on t, using (261), $2°$ using the unique factorization of m.

(346) **Exercise.** Find a complete solution of

$$x^4 - 1 \equiv 0 \pmod{30}.$$

(347) **Exercise.** Find a complete solution of

$$x^3 + x + 2 \equiv 0 \pmod{36}.$$

(348) **Exercise.** Find a complete solution of

$$x^3 + 5 \equiv 0 \pmod{60}.$$

(349) **Exercise.** Find a complete solution of the system

$$x^2 + 3 \equiv 0 \pmod{14}$$
$$3x - 6 \equiv 0 \pmod{15}.$$

34 CONGRUENCES WITH PRIME POWER MODULI

(350) Theorem (343) reduces the problem of solving $f(x) \equiv 0 \pmod{m}$ to that of solving congruences of the form $f(x) \equiv 0 \pmod{p^\alpha}$. For these, however, our only technique is the crudest—direct substitution of values from some complete residue system modulo p^α. A very simple observation may save us some work here. The observation: that if $f(x) \equiv 0 \pmod{p^\alpha}$, then $f(x) \equiv 0 \pmod{p}$. By considering the latter congruence first we may rule out certain integers as solutions of the former.

Consider, for example, $x^3 + 2 \equiv 0 \pmod{9}$. Any solution of this satisfies $x^3 + 2 \equiv 0 \pmod{3}$. Trying $x = 0, 1$, and -1 in the latter congruence gives us the complete solution $x = 1$. Thus we can restrict our search for solutions of $x^3 + 2 \equiv 0 \pmod{9}$ to the integers in some complete residue system modulo 9 congruent to 1 modulo 3; 1, 4, and 7, for example. Substitution shows that none of these work; the congruence has no solutions. [Note that it is not necessary to compute 7^3. We have $7^2 = 49 \equiv 4 \pmod{9}$, so $7^3 \equiv 7 \cdot 4 = 28 \equiv 1 \pmod{9}$. Then $7^3 + 2 \equiv 3 \not\equiv 0 \pmod{9}$.]

As another example, let us look at $f(x) = x^3 + 3 \equiv 0 \pmod{16}$. We start with $x^3 + 3 \equiv 0 \pmod{2}$; a complete solution is $x = 1$. Thus the possible solutions of $x^3 + 3 \equiv 0 \pmod{4}$ are $x = 1$ and $x = 3$. Substitution shows only $x = 1$ works. Since all solutions of $x^3 + 3 \equiv 0 \pmod{8}$ must be congruent to 1 modulo 4, we try $x = 1$ and $x = 5$ in this congruence. Since $f(1) = 4$ and $f(5) = 128$, only $x = 5$ works. We note $x = 5$ is a solution of $f(x) \equiv 0 \pmod{16}$; $x = 13$ is the only other possibility. But $13^3 + 3 \equiv (-3)^3 + 3 \equiv -27 + 3 \equiv -24 \not\equiv 0 \pmod{16}$; $x = 5$ is a complete solution to $f(x) \equiv 0 \pmod{16}$.

(351) **Exercise.** Find complete solutions to each of the following congruences.

(a) $x^2 + 6 \equiv 0 \pmod{25}$
(b) $x^2 - 3x - 4 \equiv 0 \pmod{25}$
(c) $x^3 - 2 \equiv 0 \pmod{25}$
(d) $x^3 - x \equiv 0 \pmod{16}$.

(352) **Exercise.** Let k and k' be the number of elements in complete solutions of $f(x) \equiv 0 \pmod 3$ and $f(x) \equiv 0 \pmod 9$ respectively. Construct polynomials f such that

(a) $k = 3, k' = 0$
(b) $k = 1, k' = 3$
(c) $k = 1, k' = 1$
(d) $k = 2, k' = 4$.

(353) Let us see if we can devise a general procedure for determining the solutions of $f(x) \equiv 0 \pmod{p^n}$ from those of $f(x) \equiv 0 \pmod p$. To keep things simple we start with the case $n = 2$. We are assuming f is a polynomial. Let x_1 be a solution of $f(x) \equiv 0 \pmod p$; we are interested in how x_1 generates solutions of $f(x) \equiv 0 \pmod{p^2}$. It suffices to consider the numbers $x = x_1 + yp$, $y = 0, 1, \ldots, p - 1$, since these comprise all x congruent to x_1 modulo p in a complete residue system modulo p^2. The question is for which $y, 0 \leqslant y < p$, $f(x_1 + yp) \equiv 0 \pmod{p^2}$.

Consider, for example, $x^3 + x^2 - 1 \equiv 0 \pmod{25}$. The congruence $x^3 + x^2 - 1 \equiv 0 \pmod 5$ has $x_1 = -2$ as a complete solution. Thus we seek y such that $(-2 + 5y)^3 + (-2 + 5y)^2 - 1 \equiv 0 \pmod{25}$. This says $(-2)^3 + 3(-2)^2 \cdot (5y) + 3(-2) (5y)^2 + (5y)^3 + (-2)^2 + 2(-2) (5y) + (5y)^2 - 1 \equiv 0 \pmod{25}$. Of course anything containing 5^2 as a factor is congruent to 0 modulo 25 and may be canceled. Thus our congruence simplifies to $(-2)^3 + 3(-2)^2(5y) + (-2)^2 + 2(-2) (5y) - 1 \equiv 0 \pmod{25}$, or $40y \equiv 5 \pmod{25}$. By (324) this linear congruence is equivalent to $8y \equiv 1 \pmod 5$; a complete solution $\pmod 5$ is $y = 2$. Thus a complete solution of $x^3 + x^2 - 1 \equiv 0 \pmod{25}$ is $x = -2 + 5(2) = 8$.

(354) **Exercise.** One solution of $x^3 + x^2 + 2 \equiv 0 \pmod 7$ is $x_1 = 2$. Find all solutions of the form $x = 2 + 7y, 0 \leqslant y < 7$, to $x^3 + x^2 + 2 \equiv 0 \pmod{49}$ as in the last example. Does this give a complete solution?

(355) It would be very pleasant indeed if we could always determine the solutions of $f(x) \equiv 0 \pmod{p^2}$ from those of $f(x) \equiv 0 \pmod p$ by solving a linear congruence, as in the example of (353). This would come about if in general all terms of $f(x_1 + yp) \equiv 0 \pmod{p^2}$ involving the second or higher powers of y canceled out, i.e., contained p^2 as a factor. Suppose $f(x) = a_t x^t + a_{t-1} x^{t-1} + \cdots + a_1 x + a_0$. We would like the coefficients of y^2, y^3, etc., in $a_t(x_1 + yp)^t + a_{t-1}(x_1 + yp)^{t-1} + \cdots + a_0$ to be divisible by p^2. Since we know nothing about the a's we need that each expression $(x_1 + yp)^n$, $0 \leqslant n \leqslant t$, have this property.

But according to the Binomial Theorem [see (358)] $(x_1 + yp)^n = x_1{}^n + nx_1{}^{n-1}(yp) + k_1 x_1{}^{n-2}(yp)^2 + k_2 x_1{}^{n-3}(yp)^3 + \ldots$, where the k's are integers. Of course the Binomial Theorem tells us explicitly what the k's are, but this is really more than we need. We only care that $(x_1 + yp)^n \equiv x_1{}^n + nx_1{}^{n-1} yp$ (mod p^2).

As a matter of fact, this modest result is easy to prove without invoking the full Binomial Theorem. After all, $(x_1 + yp)^n = \overbrace{(x_1 + yp)(x_1 + yp) \ldots (x_1 + yp)}^{n \text{ times}}$; if we multiply this out how can we get terms *not* containing p^2 as a factor? Clearly in two ways. We could choose x_1 from each term, forming x_1^n, or else take $n - 1$ of the x_1's and one yp. There are n ways of doing the latter, depending on from which factor we take the yp. Since this argument in no way depends on the primality of p, we have really proved

(356) **Proposition.** If n is any positive integer

$$(x + ym)^n \equiv x^n + nx^{n-1} ym \pmod{m^2}.$$

(357) **Exercise.** Prove (356) by induction on n.

(358) **Exercise*.** Prove the Binomial Theorem: If n is a positive integer, then

$$(*) \quad (a + b)^n = a^n + \frac{na^{n-1}b}{1} + \frac{n(n-1)a^{n-2}b^2}{1 \cdot 2} + \cdots +$$
$$\frac{n(n-1) \ldots (n - r + 1)a^{n-r}b^r}{1 \cdot 2 \cdot \cdots \cdot r} + \cdots .$$

[*Hint*: Assume the theorem for $n - 1$. Multiply out $(a + b) \cdot (a + b)^{n-1}$, using the expansion for the second factor. Combine terms involving the same power of a to get $(*)$.]

(359) **Exercise.** Prove that

$$1 + n/1 + n(n-1)/1 \cdot 2 + n(n-1)(n-2)/1 \cdot 2 \cdot 3 + \cdots = 2^n$$
$$1 - n/1 + n(n-1)/1 \cdot 2 - n(n-1)(n-2)/1 \cdot 2 \cdot 3 + \cdots = 0.$$

(360) **Exercise.** Since $(x + y)^n$ is clearly an integral polynomial, the Binomial Theorem shows that $n(n-1) \ldots (n - r + 1)/1 \cdot 2 \cdot \cdots \cdot r = n!/r!(n - r)!$ is an integer for $1 \leqslant r \leqslant n$. A direct proof is not so easy to come by. Criticize the following argument: The r consecutive integers $n, n - 1, \ldots, n - r + 1$ contain a complete residue system modulo k for $1 \leqslant k \leqslant r$. Thus one of them is divisible by k. Since this works for $k = 1, 2, \ldots, r$ we conclude $r! | n(n-1) \ldots (n - r + 1)$.

(361) **Exercise.** Show that if a and b are positive integers and $b = qa + r$, $0 \leqslant r < a$, then exactly q of the integers $1, 2, \ldots, b$ are divisible by a.

(362) **Exercise.** Suppose p is a prime, b is a positive integer, and $b = q_i p^i + r_i$, $0 \leqslant r_i < p^i$, for $i = 1, 2, \ldots$. Let $p^t \| b!$. Prove $t = \sum_{i=1}^{\infty} q_i$. [See (91) for the definition of $\|$.]

(363) **Exercise.** Suppose a, b, and b' are positive integers, with $b + b' = b''$. Let $b = qa + r$, $b' = q'a + r'$, and $b'' = q''a + r''$, with $0 \leqslant r, r'$, $r'' < a$. Show that $q + q' \leqslant q''$.

(364) **Exercise.** Suppose p is a prime and b and b' are positive integers, with $b + b' = b''$. Suppose $p^t \| b!$ and $p^{t'} \| b'!$. Show that $p^{t+t'} | b''!$.

(365) **Exercise.** Show that if n is a positive integer and $0 \leqslant r \leqslant n$, then $r!(n-r)! | n!$.

(366) **Exercise.** Show that $3^{48} \| 100!$.

(367) **Exercise.** Solve
   (a) $2^x \| 100!$
   (b) $7^y \| 462!$
   (c) $6^z | 45!, 6^{z+1} \nmid 45!$.

(368) **Exercise.** Prove or disprove: If p_1 and p_2 are primes, $p_1 < p_2$, and $p_2^{\alpha} | n!$, then $p_1^{\alpha} | n!$.

(369) **Exercise.** Prove that if n and r are nonnegative integers, then $r! | (n+1) \cdot (n+2) \ldots (n+r)$.

(370) **Exercise.** We defined $p^{\alpha} \| a$ only for p prime. Why?

35 CONGRUENCES WITH MODULUS p^2

(371) With the help of Proposition (356) we can say precisely what $f(x_1 + yp) \equiv 0 \pmod{p^2}$ reduces to. If $f(x) = a_t x^t + a_{t-1} x^{t-1} + \cdots + a_0$, then $f(x_1 + yp) \equiv a_t (x_1^t + t x_1^{t-1} yp) + a_{t-1} (x_1^{t-1} + (t-1)x_1^{t-2} yp) + \cdots + a_0 = f(x_1) + (t a_t x_1^{t-1} + (t-1)a_{t-1} x_1^{t-2} + \cdots + a_1) yp \pmod{p^2}$.

(372) **Definition.** Let $f(x) = a_t x^t + a_{t-1} x^{t-1} + \cdots + a_1 x + a_0$. By f' we mean the polynomial

$$t a_t x^{t-1} + (t-1) a_{t-1} x^{t-2} + \cdots + a_1.$$

We call f' the *formal derivative* of f.

(373) *Note.* Of course f' is identical with the usual derivative studied in calculus.

(374) **Example.** If $f(x) = 5x^3 - 7x^2 + 3x - 12$, then $f'(x) = 3 \cdot 5x^2 - 2 \cdot 7x + 3 = 15x^2 - 14x + 3$.

(375) We have discovered that, in our new notation, $f(x_1 + yp) \equiv f(x_1) + f'(x_1) yp \pmod{p^2}$. Recall that x_1 was a solution of $f(x) \equiv 0 \pmod{p}$; we are

trying to determine the solutions of $f(x) \equiv 0$ (mod p^2) congruent to x_1 modulo p. We must solve the linear congruence $f'(x_1) py \equiv -f(x_1)$ (mod p^2). By our assumption $p|f(x_1)$. Thus we may invoke (324) to divide through by p, getting $f'(x_1) y \equiv -f(x_1)/p$ (mod p). We are interested in $0 \leqslant y < p$; i.e., we desire a complete solution to the last congruence.

The theory for solving $f'(x_1) y \equiv -f(x_1)/p$ (mod p) has already been worked out; it is summarized in Theorem (327). First we compute $(p, f'(x_1))$. Since p is prime this is more than 1 or not according as p divides $f'(x_1)$ or not. If $p \nmid f'(x_1)$ a complete solution of $f'(x_1) y \equiv -f(x_1)/p$ (mod p) contains 1 element. It can be found either by trial and error or by the Euclidean Algorithm method, depending on the size of p.

If $p | f'(x_1)$ there is no solution unless p also divides $-f(x_1)/p$. In the latter case any value of y makes the congruence true. Let us put together what we have figured out.

(376) **Proposition.** Consider $f(x) \equiv 0$ (mod p^2). Suppose x_1 is a solution of $f(x) \equiv 0$ (mod p). Then x_1 generates 0, 1, or p solutions to the first congruence of the form $x_1 + py$, according as a complete solution of $f'(x_1)y \equiv -f(x_1)/p$ (mod p) has 0, 1, or p elements.

(377) *Note.* Of course if $f(x) \equiv 0$ (mod p) has two or more solutions the above technique must be applied to each of them in order to find a complete solution of $f(x) \equiv 0$ (mod p^2).

(378) **Example.** Consider $f(x) = x^3 + x^2 + 3 \equiv 0$ (mod 25). The congruence $f(x) \equiv 0$ (mod 5) has a complete solution $x_1 = 1, 2$. (It is easiest to substitute from the complete residue system $-2, -1, 0, 1, 2$ into f.)

First we work with $x_1 = 1$. Since $f'(x) = 3x^2 + 2x$ we have $f(x_1) = 5$, $f'(x_1) = 5$. We must solve

$$5y \equiv -5/5 \text{ (mod 5)}.$$

There are no solutions.

Now we try $x_1 = 2$. Here $f(x_1) = 15, f'(x_1) = 16$. We must solve

$$16y \equiv -15/5 \text{ (mod 5), or}$$

$$y \equiv -3 \text{ (mod 5)}.$$

This has the complete solution $y = 2$. We find $x = 2 + 2 \cdot 5 = 12$ is a complete solution to $x^3 + x^2 + 3 \equiv 0$ (mod 25).

(379) **Exercise.** Solve $x^3 + x^2 \equiv 0$ (mod 25).

(380) **Exercise.** Solve $x^3 + x^2 + 4 \equiv 0$ (mod 49).

(381) **Exercise.** Solve $x^4 - 1 \equiv 0$ (mod 25).

(382) **Exercise.** Solve $x^4 - 1 \equiv 0$ (mod 49).

(383) **Exercise.** Solve $x^2 + 8 \equiv 0$ (mod 121).

36 CONGRUENCES WITH MODULUS p^n

(384) Solutions of $f(x) \equiv 0$ (mod p^2) are generated from those of $f(x) \equiv 0$ (mod p) by the solution of a linear congruence modulo p. Do these in turn generate solutions to $f(x) \equiv 0$ (mod p^3)? More generally, do solutions of $f(x) \equiv 0$ (mod p^n) generate solutions of $f(x) \equiv 0$ (mod p^{n+1}) in some similarly simple way? If the answer is yes, we could start by solving the congruence modulo p and build up to complete solutions modulo any power of p.

Suppose x_n is a solution to $f(x) \equiv 0$ (mod p^n). If y runs through a complete residue system modulo p, then $x_n + yp^n$ runs through all elements of some complete residue system modulo p^{n+1} congruent to x_n modulo p^n. (It is perhaps well to remind ourselves at this point that $f(x) \equiv 0$ (mod p^{n+1}) implies $f(x) \equiv 0$ (mod p^n); thus all solutions to the former congruence are congruent modulo p^n to solutions of the latter.) When is $f(x_n + yp^n) \equiv 0$ (mod p^{n+1})? As before, let $f(x) = a_t x^t + \cdots + a_0$. We note that

$$(x_n + yp^n)^r = x_n^r + rx_n^{r-1}(yp^n) + \text{ a multiple of } (yp^n)^2 .$$

Since $2n \geqslant n + 1$ for $n \geqslant 1$, we see

$$(x_n + yp^n)^r \equiv x_n^r + rx_n^{r-1}yp^n \pmod{p^{n+1}}$$

for all r. Thus

$$f(x_n + yp^n) \equiv a_t(x_n^t + tx_n^{t-1}yp^n) + \cdots + a_0$$
$$= f(x_n) + f'(x_n)yp^n \pmod{p^{n+1}}.$$

It suffices to solve $f'(x_n)yp^n \equiv -f(x_n)$ (mod p^{n+1}). Since $p^n | f(x_n)$ by assumption, this is equivalent to

$$f'(x_n)y \equiv -f(x_n)/p^n \pmod{p}.$$

As before, this congruence will have 0, 1, or p solutions: 0 if $p|f'(x_n)$ but $p^{n+1} \nmid f(x_n)$, 1 if $p \nmid f'(x_n)$, and p if $p|f'(x_n)$ and $p^{n+1}|f(x_n)$.

(385) **Example.** Consider $x^3 + x^2 + 23 \equiv 0$ (mod 125). First we find a complete solution of $f(x) = x^3 + x^2 + 23 \equiv 0$ (mod 5) by trial and error; one is $x_1 = 1, 2$. We compute $f'(x) = 3x^2 + 2x$. We must solve

$$f'(x_1)y \equiv -f(x_1)/5 \pmod 5.$$

For $x_1 = 1$, this is

$$5y \equiv - {}^{25}\!/_5 \pmod 5.$$

Here $y = 0, \pm 1, \pm 2$ is a complete solution. Thus we have generated the solutions $x_2 = 1 + 5y = 1, 6, 11, -4, -9$.

In the same way taking $x_1 = 2$ leads to the congruence

$$16y \equiv - {}^{35}\!/_5 \pmod 5.$$

A complete solution is $y = -2$, giving $x_2 = 2 + 5(-2) = -8$. We have found a complete solution of $f(x) \equiv 0$ (mod 25): $x_2 = 1, 6, 11, -4, -9, -8$.

We consider these in turn. It is convenient to make a little table of f and f' so as not to have to repeat calculations.

x	$f(x)$	$f'(x)$
0	23	
1	25	5
-1	23	
2	35	16
-2	19	
6	275	
11	1475	
-4	-25	$\equiv 5$ (mod 5)
-9	-625	
-8	-425	$\equiv 16$ (mod 5)

Notice that all we ever need to know about $f'(x_2)$ is what it is modulo 5. This will be the same among all solutions generated by a given x_1, since all are congruent modulo 5. Thus since $f'(1) \equiv 0$ (mod 5), the same is true for $f'(x_2)$ for $x_2 = 6, 11, -4$, and -9. For these (and for $x_2 = 1$) we must solve

$$0 \cdot y \equiv -f(x_2)/25 \text{ (mod 5)}.$$

Inspection of our table reveals this is solvable only for $x_2 = -9$; in this case $y = 0, \pm 1, \pm 2$ is a complete solution. We have generated the solutions $x_3 = -9 + 25y = -9, 16, 41, -34, -59$.

We turn to $x_2 = -8$. Here we must solve

$$16y \equiv 425/25 \text{ (mod 5)}.$$

A complete solution is $y = 2$. This gives $x_3 = -8 + 2 \cdot 25 = 42$.

A complete solution to $x^3 + x^2 + 23 \equiv 0$ (mod 125) is thus $x_3 = -9, 16, 41, -34, -59, 42$.

(386) **THEOREM.** Suppose $x_n, x_n', x_n'', \ldots$ is a complete solution to $f(x) \equiv 0$ (mod p^n), where f is a polynomial and p is a prime. All solutions of $f(x) \equiv 0$ (mod p^{n+1}) congruent to x_n modulo p^n are given by $x_{n+1} = x_n + yp^n$, where y runs through any complete solution of

$$f'(x_n)y \equiv -f(x_n)/p^n \text{ (mod } p).$$

Applying this also to $x_n', x_n'', \ldots$ in turn yields a complete solution to $f(x) \equiv 0$ (mod p^{n+1}).

(387) **Exercise.** Solve $x^2 - x + 7 \equiv 0$ (mod 27).

(388) **Exercise.** Solve $x^2 - 44 \equiv 0$ (mod 81).

(389) **Exercise.** Solve $x^2 + 3x - 50 \equiv 0$ (mod 125).

(390) **Exercise.** Solve $x^3 + x^2 + 10 \equiv 0 \pmod{32}$.

(391) **Exercise.** Solve $x^3 + x^2 + 1 \equiv 0 \pmod{27}$.

(392) **Exercise.** If $f(x_0) = 0$, clearly $f(x_0) \equiv 0 \pmod{p^n}$ for $n = 1, 2, \ldots$. Suppose $f(x_0) \equiv 0 \pmod{p^n}$, $n = 1, 2, \ldots$, p a prime and f a polynomial. Prove $f(x_0) = 0$.

(393) **Exercise.** Let S be a finite set, f a polynomial, and p a prime. Suppose given n there exists x_n in S such that $f(x_n) \equiv 0 \pmod{p^n}$. Prove there exists x_0 in S such that $f(x_0) = 0$.

(394) **Exercise.** Prove or disprove: If $f(x) \equiv 0 \pmod{p^n}$ has a solution for $n = 1, 2, \ldots$, f a polynomial and p a prime, then there exists an integer x_0 such that $f(x_0) = 0$.

(395) **Exercise.** Suppose p is a prime and f a polynomial such that $p \mid f(x_0)$ but $p \nmid f'(x_0)$. Show that $f(x) \equiv 0 \pmod{p^n}$ has a solution for $n = 1, 2, \ldots$.

(396) **Exercise.** Let h be the number of elements in a complete solution of $f(x) \equiv 0 \pmod{p^2}$, where f is a polynomial and p a prime. Suppose $p \nmid h$. Show $f(x) \equiv 0 \pmod{p^n}$ has a solution for $n = 1, 2, \ldots$.

(397) **Exercise.** For what integers h does there exist a polynomial f such that a complete solution of $f(x) \equiv 0 \pmod 9$ has exactly h elements?

(398) **Exercise.** Solve $x^2 - x + 12 \equiv 0 \pmod{1000}$.

(399) **Exercise.** Solve $10x^2 - x^3 \equiv 9 \pmod{405}$.

(400) **Exercise.** Solve the system

$$x^2 - 3x + 6 \equiv 0 \pmod 8$$
$$x^3 + x^2 + 3 \equiv 0 \pmod 9.$$

(401) **Exercise.** Solve the system

$$x^2 - 9 \equiv 0 \pmod{16}$$
$$x^3 - 10 \equiv 0 \pmod{18}.$$

37 CONGRUENCES OF DEGREE 2

(402)   The Chinese Remainder Theorem reduces solving the polynomial congruence $f(x) \equiv 0 \pmod m$ to solving congruences of the form $f(x) \equiv 0 \pmod{p^n}$, where p is prime. Our last theorem reduces this to solving $f(x) \equiv 0 \pmod p$, plus an undetermined number of linear congruences. Since the latter are covered by Theorem (327), the only gap in our theory comes in solving congruences to

prime moduli; here we must still resort to trial and error. This is not bad for small primes like 5 or 7, but quickly becomes tedious as p increases. Substituting a complete residue system modulo 31 into so simple a polynomial congruence as

$$x^2 + 6x + 1 \equiv 0 \pmod{31},$$

for example, is more work than most people would like to do without getting paid for it.

Of course we have a method for solving *some* polynomial congruences with modulus p, namely the linear ones. Perhaps we should see what we can do at the next step up, congruences involving polynomials of degree 2. Let us consider $ax^2 + bx + c \equiv 0 \pmod{p}$.

If this were an equation instead of a congruence we could solve it quite readily by means of the quadratic formula

$$x = \frac{-b \pm \sqrt{b^2 - 4ac}}{2a}.$$

Perhaps we can work this into a solution of the congruence. In what follows we will give an example of a type of reasoning, compounded of intuition, guess-work, and wishful thinking, that pervades mathematical creation yet almost never gets into print. Although sometimes completely fruitless, such uncritical thinking often carries one close enough to the truth to direct the application of more rigorous arguments, just as an illegal wiretap, while not admissible in court, may lead the police to construct a legitimate case.

Can we make any sense out of

$$x \equiv \frac{-b \pm \sqrt{b^2 - 4ac}}{2a} \pmod{p}?$$

The prospect is not rosy; the right-hand side may not even be an integer. One problem is the division by $2a$. Let us write

$$2ax \equiv -b \pm \sqrt{b^2 - 4ac} \pmod{p}$$

instead. If the right-hand side of this is an integer, a solution x is assured as long as $(2a, p) = 1$, true unless $p = 2$ or $p|a$. We should be so lucky as to have $p = 2$. Congruences modulo 2 give us no trouble; let us assume $p \neq 2$. If $p|a$, then $ax^2 \equiv 0 \pmod{p}$ no matter what x is, so our original congruence is equivalent to $bx + c \equiv 0 \pmod{p}$. Since we have already handled such linear congruences, we assume $p \nmid a$.

This is all fine so long as $\sqrt{b^2 - 4ac}$ is an integer, but this seems much too much to hope for. Perhaps it suffices that $b^2 - 4ac$ be a perfect square modulo p; i.e., there exists an integer y such that

$$y^2 \equiv b^2 - 4ac \pmod{p}.$$

It is about time to come down out of the clouds. We think that if y is a solution of $y^2 \equiv b^2 - 4ac \pmod{p}$, and if x then is a solution of $2ax \equiv -b + y$

(mod p), then x *might* be a solution of $ax^2 + bx + c \equiv 0$ (mod p). Let us try an example to see if we have anything, say $2x^2 + 3x + 3 \equiv 0$ (mod 5). Here $a = 2$, $b = 3$, $c = 3$, and $b^2 - 4ac = 9 - 24 = -15$. The congruence $y^2 \equiv -15$ (mod 5) clearly has $y = 0$ as a complete solution. Next we solve $2ax \equiv -b + y$ (mod p), or

$$4x \equiv -3 \text{ (mod 5)}.$$

This has $x = 3$ as a complete solution. Is $x = 3$ a solution of the original congruence? It is; in fact trial and error shows $x = 3$ is a complete solution.

As another example, let us consider

$$x^2 + 6x + 1 \equiv 0 \text{ (mod 31)}.$$

Here $b^2 - 4ac = 36 - 4 = 32$. One solution of $y^2 \equiv 32 \equiv 1$ (mod 31) is $y = 1$. Solving $2x \equiv -6 + 1 \equiv 26$ (mod 31) gives $x = 13$ as a solution. We note that $13^2 + 6 \cdot 13 + 1 = 248 = 8 \cdot 31$. It works.

Notice that $(-1)^2 \equiv 32$ (mod 31) also. Solving $2x \equiv -6 - 1 \equiv 24$ (mod 31) gives $x = 12$. Note that $12^2 + 6 \cdot 12 + 1 = 217 = 7 \cdot 31$.

Whether $x = 12, 13$ is a *complete* solution of $x^2 + 6x + 1 \equiv 0$ (mod 31) remains to be seen, but at least we have found two solutions.

(403)   The above musing seems promising, but of course a number of questions remain to be answered:

1. Given that $y^2 \equiv b^2 - 4ac$ (mod p) and $2ax \equiv -b + y$ (mod p), is x necessarily a solution of $ax^2 + bx + c \equiv 0$ (mod p)?

2. Are all solutions of this form?

3. How many elements may we expect a complete solution of $ax^2 + bx + c \equiv 0$ (mod p) to have?

We consider (1) first. The best way to check a candidate for a solution is direct substitution. The dirty work is assigned to the reader.

(404)   **Exercise***. Suppose $(2a,p) = 1$, $y^2 \equiv b^2 - 4ac$ (mod p), and $2ax \equiv -b + y$ (mod p). Show $ax^2 + bx + c \equiv 0$ (mod p). [*Hint*: There exists an integer z such that $2az \equiv 1$ (mod p). Then $x \equiv z(-b + y)$ (mod p). Substitute this in $ax^2 + bx + c$.]

(405)   Now we turn to answering (2). The reader may remember that the proof of the quadratic formula consists of two parts: (1) showing $(-b + \sqrt{b^2 - 4ac})/2a$ is a solution, and (2) showing every solution has this form. The latter may be accomplished by assuming $ax^2 + bx + c = 0$ and completing the square: $x^2 + bx/a + c/a = 0$, so $x^2 + bx/a + b^2/4a^2 - b^2/4a^2 + c/a = (x + b/2a)^2 + c/a - b^2/4a^2 = 0$, so $(x + b/2a)^2 = b^2/4a^2 - c/a$, etc. Not being too proud to copy a good idea, let us try the same thing on $ax^2 + bx + c \equiv 0$ (mod p). Of course we must try to arrange things so as to keep everything integral; otherwise the congruence wouldn't make sense. Since this means we cannot divide through by a we will multiply by it; this also has the effect of making our leading term a perfect square:

$$a^2x^2 + abx + ac \equiv 0 \pmod{p}.$$

We would like to write this as

$$(ax + b/2)^2 - b^2/4 + ac \equiv 0 \pmod{p},$$

but maybe $4\nmid b^2$. Multiplying by 4 first gets around this:

$$4a^2x^2 + 4abx + 4ac \equiv 0 \pmod{p},$$

or

$$(2ax + b)^2 - b^2 + 4ac \equiv 0 \pmod{p}.$$

We see $b^2 - 4ac \equiv (2ax + b)^2 \pmod{p}$. If we call $2ax + b = y$, we have $2ax \equiv -b + y \pmod{p}$ and $y^2 \equiv b^2 - 4ac \pmod{p}$. This proves (2).

(406) Answering (3) should not be difficult since our previous results tell us exactly what all solutions of $ax^2 + bx + c \equiv 0 \pmod{p}$ look like. They are the x such that $2ax \equiv -b + y \pmod{p}$, where $y^2 \equiv b^2 - 4ac \pmod{p}$. Under the assumption that $(2a,p) = 1$, the linear congruence $2ax \equiv -b + y \pmod{p}$ has a unique solution modulo p for each value of y; thus it suffices to determine the number of elements in a complete solution of the congruence $y^2 \equiv b^2 - 4ac \pmod{p}$. There may be no solutions at all; for example $y^2 \equiv 2 \pmod{3}$ has no solution. On the other hand, if y is a solution, so is $-y$. Can there be more than 2 solutions? Suppose y_0 is a solution; then the congruence is equivalent to $y^2 \equiv y_0^2 \pmod{p}$. This says $p|y^2 - y_0^2 = (y - y_0)(y + y_0)$. We know that since p is prime $p|y - y_0$ or $p|y + y_0$. This says $y \equiv \pm y_0 \pmod{p}$. Thus we seem to have proved

(407) **False Statement.** If p is a prime and $(2a,p) = 1$, then the congruence $ax^2 + bx + c \equiv 0 \pmod{p}$ has either 0 or 2 elements in any complete solution, depending on whether $y^2 \equiv b^2 - 4ac \pmod{p}$ is solvable or not.

(408) If he hasn't already, the reader is invited to decide why we haven't proved (407). Here are a few exercises to work on while deciding.

(409) **Exercise.** Solve $4x^2 + 6x + 1 \equiv 0 \pmod{19}$.

(410) **Exercise.** Solve $4x^2 + 6x + 1 \equiv 0 \pmod{11}$.

(411) **Exercise.** Solve $4x^2 + 6x + 1 \equiv 0 \pmod{13}$.

(412) **Exercise.** Solve $x^2 + 16x + 32 \equiv 0 \pmod{127}$.

(413) **Exercise.** Solve $5x^2 + 8x + 1 \equiv 0 \pmod{129}$.

(414) **Exercise.** Solve $3x^2 + 3x - 5 \equiv 0 \pmod{53}$.

(415) The only thing wrong with the analysis given in (406) is that we slipped into the assumption that if y_0 is a solution of $y^2 \equiv b^2 - 4ac \pmod{p}$, then $-y_0$ is *another* element of some complete solution. What if $-y_0 \equiv y_0 \pmod{p}$? This says $2y_0 \equiv 0 \pmod{p}$, or $p|2y_0$. We are assuming $p \neq 2$, so $p|y_0$. Since this

happens if and only if $p \mid y_0^2$, the congruence $y^2 \equiv b^2 - 4ac$ has exactly one solution just in the case that $p \mid b^2 - 4ac$. The first example we worked through in (402) illustrated such a case.

(416) **THEOREM.** The solutions x of the congruence

$$ax^2 + bx + c \equiv 0 \ (\mathrm{mod}\ p),$$

where p is a prime and $(2a,p) = 1$, are given by the solutions of $2ax \equiv -b + y$ (mod p), where y runs through a complete solution of $y^2 \equiv b^2 - 4ac$ (mod p). A complete solution has 1 element if $p \mid b^2 - 4ac$, otherwise 0 or 2 elements.

38 QUADRATIC RESIDUES

(417) The problem of solving $y^2 \equiv b^2 - 4ac$ (mod p) remains. The question is that of identifying the integers that are squares modulo p.

(418) **Definition.** Let m be a positive integer. We say the integer a is a *quadratic residue modulo m* in case $(a,m) = 1$ and there exists an integer x such that $x^2 \equiv a$ (mod m).

(419) **Example.** Let $m = 6$. We have $0^2 \equiv 0$, $1^2 \equiv 1$, $2^2 \equiv 4$, $3^2 = 9 \equiv 3$, $4^2 = 16 \equiv 4$, and $5^2 = 25 \equiv 1$ (mod 6). Since if $x^2 \equiv a$ (mod 6) is solvable at all it has a solution in any complete residue system (mod 6), we see that the congruence has a solution for $a \equiv 0$, 1, 3, and 4 (mod 6) but not for $a \equiv 2$ or 5 (mod 6). In light of the condition $(a,m) = 1$, we see that among the integers from 0 to 5 only the integer 1 is a quadratic residue.

The stipulation in our definition that a be relatively prime to m in order to be a quadratic residue may seem unnatural to the reader, but the definition given is the standard one. It turns out that the determination of the quadratic residues in a reduced residue system is more easily treated than in a complete residue system [see (426), for example]. In what follows our main attention will be directed toward prime moduli, when the case $a = 0$ is trivial.

(420) **Exercise.** Determine all integers a, $0 \leqslant a \leqslant 18$, such that a is a quadratic residue modulo 19.

(421) **Exercise.** Determine all integers a, $0 \leqslant a \leqslant 19$, such that a is a quadratic residue modulo 20.

(422) **Exercise.** Suppose p is an odd prime. Show that if a is a quadratic residue (mod p) then a is a quadratic residue (mod p^n) for each positive integer n. [*Hint*: Apply (386).]

(423) **Exercise.** Suppose $2 \nmid a$. Show that a is a quadratic residue modulo 2, while a is a quadratic residue modulo 4 if and only if $a \equiv 1$ (mod 4).

(424) **Exercise.** Use (386) to solve $x^2 \equiv 17 \pmod{64}$.

(425) **Exercise.** Suppose $n \geqslant 3$. Show that a is a quadratic residue modulo 2^n if and only if $a \equiv 1 \pmod 8$. [*Hint*: The "if" part is harder. The congruence $x^2 \equiv 1 \pmod 8$ has 4 solutions, 2 of which satisfy $x^2 \equiv a \pmod{16}$ and 2 of which do not. Use (386) to show $x^2 \equiv a \pmod{16}$ also has exactly 4 solutions, half satisfying $x^2 \equiv a \pmod{32}$. Proceed by induction.]

(426) **Exercise.** Show a is a quadratic residue modulo m if and only if a is a quadratic residue modulo p for each odd prime p dividing m and $a \equiv 1 \pmod{2^k}$ if $2^k \mid m, k = 2, 3$.

(427) Although we have given a quite general definition of quadratic residue, we are mainly interested in determining whether a is a quadratic residue modulo p, where p is an odd prime and $p \nmid a$. Theorem (416) (along with previous results) reduces solving arbitrary congruences of degree 2 to this problem.

If $p = 11$, for example, we wonder which elements in some reduced residue system are quadratic residues and which are not. (Elements of the second sort are called "nonresidues.") Since if $p \nmid a$ and $x^2 \equiv a \pmod p$, then $p \nmid x$ also, it suffices to compute $x^2 \pmod p$ for $x = 1, 2, \ldots, 10$. We have

x	1	2	3	4	5	6	7	8	9	10
x^2	1	4	9	$16 \equiv 5$	$25 \equiv 3$	$36 \equiv 3$	$49 \equiv 5$	$64 \equiv 9$	$81 \equiv 4$	$100 \equiv 1$

all the congruences being modulo 11. Thus we have 5 residues, 1, 3, 4, 5, and 9, and 5 nonresidues modulo 11 in our set. We see from our table that the 5 quadratic residues form the first 5 entires, then recur in opposite order. This is because $x^2 = (-x)^2 \equiv (11 - x)^2 \pmod{11}$ for all x. In general $x^2 \equiv (p - x)^2 \pmod p$, so a complete set of quadratic residues modulo p will always be found among the first half of the integers $1^2, 2^2, \ldots, (p - 1)^2$; namely, among $1^2, 2^2, \ldots, \left(\frac{p-1}{2}\right)^2$. But we saw in (406) that if a is a quadratic residue modulo p, then a complete solution to $x^2 \equiv a \pmod p$ has just 2 elements. Thus the integers $1^2, 2^2, \ldots, \left(\frac{p-1}{2}\right)^2$ are incongruent modulo p.

(428) **Proposition.** Let p be any odd prime. Then any reduced residue system modulo p contains exactly $(p-1)/2$ quadratic residues and $(p-1)/2$ nonresidues modulo p. One set of $(p-1)/2$ incongruent residues is $1^2, 2^2, \ldots, \left(\frac{p-1}{2}\right)^2$.

chapter 5

The Distribution of Powers into Congruence Classes

That the numbers a^k follow a definite pattern as k runs through the positive integers will be shown in this chapter, which contains several of the most famous theorems of number theory.

39 THE ORDER OF AN INTEGER MODULO m

(429) What about higher powers of the elements of a reduced residue system modulo p? Let us make a table for $p = 7$.

a	1	2	3	4	5	6
a^2	1	4	$9 \equiv 2$	$16 \equiv 2$	$25 \equiv 4$	$36 \equiv 1$
a^3	1	$2 \cdot 4 \equiv 1$	$3 \cdot 2 \equiv 6$	$4 \cdot 2 \equiv 1$	$5 \cdot 4 \equiv 6$	$6 \cdot 1 \equiv 6$
a^4	1	$2 \cdot 1 \equiv 2$	$3 \cdot 6 \equiv 4$	$4 \cdot 1 \equiv 4$	$5 \cdot 6 \equiv 2$	$6 \cdot 6 \equiv 1$
a^5	1	$2 \cdot 2 \equiv 4$	$3 \cdot 4 \equiv 5$	$4 \cdot 4 \equiv 2$	$5 \cdot 2 \equiv 3$	$6 \cdot 1 \equiv 6$
a^6	1	$2 \cdot 4 \equiv 1$	$3 \cdot 5 \equiv 1$	$4 \cdot 2 \equiv 1$	$5 \cdot 3 \equiv 1$	$6 \cdot 6 \equiv 1$

Notice that it is not necessary to compute, for example, $3^3 = 27$, since $3^2 \equiv 2$ (mod 7) implies $3^3 \equiv 3 \cdot 2 \equiv 6$ (mod 7). Thus a table such as the above may be constructed by working down the columns. We stopped our table at a^6 because we found $a^6 \equiv 1$ (mod 7) for $a = 1, 2, \ldots, 6$; the columns clearly repeat after this point. (For $a^7 = a \cdot a^6 \equiv a$ (mod 7), $a^8 = a^2 \cdot a^6 \equiv a^2$ (mod 7), etc.) The integer 1 appears in some columns before the 6th power; namely, $2^3 \equiv 1$, $4^3 \equiv 1$, and $6^2 \equiv 1$ (mod 7). Our interest in the first appearance of 1 in a column stems from the fact that the column repeats thereafter.

We may ask whether 1 invariably appears in each column of such a table. Let m be a positive integer and suppose $(a,m) = 1$. We consider $a, a^2, a^3, \ldots$. Since there are only a finite number of congruence classes modulo m these cannot all be incongruent; there must be i and j such that $a^i \equiv a^j$ (mod m). Suppose $i < j$. Then $a^i \equiv a^i \cdot a^{j-i}$ (mod m), or $m | a^i (1 - a^{j-i})$. Since $(m, a^i) = 1$, we have $m | 1 - a^{j-i}$ (see (249)), or $a^{j-i} \equiv 1$ (mod m). Notice that in effect we canceled a^i from both sides of the congruence $a^i \equiv a^j$ (mod m). Such cancellation comes up often enough to deserve statement as a proposition, the proof of which the reader should supply.

(430) **Proposition.** If $kx_1 \equiv kx_2 \pmod{m}$ and $(k,m) = 1$, then $x_1 \equiv x_2 \pmod{m}$.

(431) **Definition.** Let m be a positive integer and suppose $(a,m) = 1$. Let g be the least positive integer such that $a^g \equiv 1 \pmod{m}$. We call g the *order of a modulo m*.

(432) *Note.* The existence of g is guaranteed by the argument at the end of (429). Other terminologies are common; e.g., "a belongs to $g \pmod{m}$," or "a belongs to the exponent g modulo m." The one we use is consistent with the language of modern algebra.

(433) **Examples.** The orders of 1, 2, 3, 4, 5, and 6 modulo 7 are, respectively, 1, 3, 6, 3, 6, and 2.

(434) **Exercise.** Find the orders of 1, 3, 7, and 9 modulo 10. Find the orders of 1, 2, . . . , 10 modulo 11.

(435) **Exercise.** Suppose $(a,m) = 1$ and n is a positive integer. Show that the order of $a^n \pmod{m}$ does not exceed the order of $a \pmod{m}$.

(436) In (429) we saw that $a^i \equiv a^j \pmod{m}$ implies $a^{j-i} \equiv 1 \pmod{m}$, where $(a,m) = 1$ and $j > i$. If a^j represents the first repetition $\pmod{m}$ in the sequence $a, a^2, a^3, \ldots$, we see that a^j is preceded by a power congruent to 1. Since if g is the order of a modulo m, a^g is the first of the integers $a, a^2, \ldots$ congruent to 1, the integers $a, a^2, \ldots, a^g$ must be incongruent modulo m. After a^g, the cycle repeats; $a^{g+1} = a^g \cdot a \equiv a \pmod{m}$, $a^{g+2} = a^g \cdot a^2 \equiv a^2 \pmod{m}$, etc. The residue 1 returns at $a^{2g} = (a^g)^2 \equiv 1^2$, and again at a^{3g}, a^{4g}, etc.

(437) **Proposition.** Suppose m is a positive integer and $(a,m) = 1$. Let the order of a modulo m be g. Then the integers $a, a^2, \ldots, a^g$ are incongruent modulo m and every positive power of a is congruent to one of them. If n is a positive integer, $a^n \equiv 1 \pmod{m}$ if and only if $g \mid n$.

40 EULER'S AND FERMAT'S THEOREMS

(438) Since $(a,m) = 1$ implies $(a^n,m) = 1$, the integers $a, a^2, \ldots, a^g$ form a subset of some reduced residue system modulo m. We have seen they are incongruent; thus $g \leqslant \varphi(m)$, the total number of elements in any reduced residue system. If $m = 7$, for example, $\varphi(m) = 6$, and we found earlier that if $7 \nmid a$, then a has order 1, 2, 3, or 6 $\pmod{7}$. In this case not only is $g \leqslant \varphi(m)$, but $g \mid \varphi(m)$ in each case.

Let us try another example, say $m = 15$.

a	1	2	4	7	8	11	13	14
a^2		4	1	4	4	1	4	1
a^3		8		13	2		7	
a^4		1		1	1		1	
g	1	4	2	4	4	2	4	2

Here we find elements of orders 1, 2, and 4, while $\varphi(15) = 8$. Again $g|\varphi(m)$ in each case.

(439) **Exercise.** Compute the order g of a modulo 13 for $1 \leqslant a \leqslant 12, (a,13) = 1$. Does $g|\varphi(13)$ in each case?

(440) It seems to be a reasonable conjecture that, in general, if g is the order of a modulo m, then $g|\varphi(m)$. By (437) this will be true if and only if $a^{\varphi(m)} \equiv 1$ for all a such that $(a,m) = 1$, which expresses our conjecture somewhat more concisely. We saw, for example, that $a^6 \equiv 1 \pmod 7$ whenever $7 \nmid a$.

Finding a proof is something else. Why should $a^{\varphi(m)}$ be congruent to 1? What does $\varphi(m)$ have to do with anything? Since $\varphi(m)$ counts the elements of a reduced residue system modulo m, maybe we should try to bring the latter into the problem somehow. Let $b_1, b_2, \ldots, b_t$ be a reduced residue system modulo m, where $t = \varphi(m)$. We have the same number of a's in $a^{\varphi(m)}$; perhaps we should match them up somehow. For example $a^{\varphi(m)}b_1 b_2 \ldots b_t = (ab_1) \cdot (ab_2) \ldots (ab_t)$. But the elements $ab_1, ab_2, \ldots, ab_t$ again form a reduced residue system $\pmod m$; we proved this in Theorem (251). Thus they are congruent to $b_1, b_2, \ldots, b_t$ in some order. In particular,

$$a^{\varphi(m)}b_1 b_2 \ldots b_t = (ab_1)(ab_2) \ldots (ab_t) \equiv b_1 b_2 \ldots b_t \pmod m.$$

This is a perfect setup for our cancellation law, Proposition (430). Since $(b_i, m) = 1$, $1 \leqslant i \leqslant t$, we have $(b_1 b_2 \ldots b_t, m) = 1$. (This is (287).) Thus we see $a^{\varphi(m)} \equiv 1 \pmod m$.

(441) **Euler's Theorem.** If m is a positive integer and $(a,m) = 1$, then $a^{\varphi(m)} \equiv 1 \pmod m$.

(442) **Fermat's Theorem.** If p is prime, then $a^p \equiv a \pmod p$. (Fermat's Theorem is really just a special case of Euler's Theorem. We list it separately since it was proved first and is well-known by this name.)

Proof. If $p|a$ the result is clear. If $p \nmid a$, then $a^{\varphi(p)} = a^{p-1} \equiv 1 \pmod p$ by Euler's Theorem. The result follows by multiplying through by a.

(443) **Exercise.** Solve $3^{100} \equiv x \pmod{17}$, $1 \leqslant x < 17$. [*Hint:* $3^{100} = 3^{6 \cdot 16 + 4} \equiv 3^4 \pmod{17}$, since $3^{16} \equiv 1 \pmod{17}$.]

(444) **Exercise.** Solve $7^{50} \equiv x \pmod{30}$, $0 < x \leqslant 30$.

(445) **Exercise.** Solve $5^{50} \equiv x \pmod{30}$, $0 \leqslant x < 30$. [*Hint:* Use the Chinese Remainder Theorem.]

(446) **Exercise.** Solve $x^{50} \equiv 2 \pmod 7$. [*Hint*: Clearly $7 \nmid x$. Thus $x^{50} = x^{6 \cdot 8 + 2} \equiv x^2 \pmod 7$.]

(447) **Exercise.** Solve $x^{60} + x^{50} + 3x^{25} \equiv 0 \pmod{13}$.

41 REDUCING THE DEGREE OF POLYNOMIAL CONGRUENCES

(448) The last two exercises show how Euler's Theorem may be used to solve a polynomial congruence by replacing high powers of the unknown with lower ones. Suppose a term involving x^k occurs in a polynomial congruence with prime modulus p. If $p \nmid x$, then $x^{p-1} \equiv 1 \pmod p$ by Euler's Theorem. We use the Euclidean Algorithm to write $k = (p-1)q + r$, where $0 \le r < p-1$. Then $x^k = (x^{p-1})q_x{}^r \equiv x^r \pmod p$. Thus our congruence can be reduced to one of degree $p-2$ or less. To be sure, the above reduction will be invalid when $p \mid x$, but whether 0 is a solution of the original congruence or not is easily tested.

(449) **THEOREM.** If p is a prime and $P(x)$ is a polynomial, then there exists a polynomial $P_0(x)$ of degree not exceeding $p-2$ such that $P(x) \equiv 0 \pmod p$ and $P_0(x) \equiv 0 \pmod p$ have the same solutions, except perhaps for $x \equiv 0 \pmod p$.

(450) **Exercise.** If $P(x)$ is a polynomial and p is a prime, then 0 is a solution of $P(x) \equiv 0 \pmod p$ if and only if p divides the constant term of $P(x)$.

(451) **Exercise.** Show that (449) may not be true if p is not prime.

(452) **Example.** Let us consider

$$P(x) = x^{24} + 3x^{19} - x^{18} + x^7 + x^2 + 3 \equiv 0 \pmod 5.$$

Here $x^{24} = x^{4 \cdot 6} \equiv 1$, $x^{19} = x^{4 \cdot 4 + 3} \equiv x^3$, $x^{18} = x^{4 \cdot 4 + 2} \equiv x^2$, and $x^7 = x^{4 \cdot 1 + 3} \equiv x^3 \pmod 5$. We look at

$$1 + 3x^3 - x^2 + x^3 + x^2 + 3 = 4x^3 + 4 \equiv 0 \pmod 5.$$

By (430) this is equivalent to $x^3 + 1 \equiv 0 \pmod 5$. A complete solution is $x = -1$. Since $P(0) \not\equiv 0 \pmod 5$, $x = -1$ is a complete solution to the original congruence.

(453) **Exercise.** Solve $x^{25} + x^{24} + 2x^{23} \equiv 0 \pmod 7$.

(454) **Exercise.** Solve $2x^{33} + 7x^{23} + 2x^3 + 3 \equiv 0 \pmod{11}$.

(455) **Exercise.** Solve $x^{24} + x^2 - 1 \equiv 0 \pmod{13}$.

42 WILSON'S THEOREM

(456) The argument we used to prove Euler's Theorem in (440) is too good to let alone. Recall we noticed that as x ran through a reduced residue system,

say R, modulo m, so did ax, where $(a,m) = 1$. Thus

$$a^{\varphi(m)} \prod_{x \in R} x = \prod_{x \in R} ax \equiv \prod_{x \in R} x \pmod{m}.$$

Cancelling $\prod_{x \in R} x$ from both sides produced the theorem.

Matching things up is the germ of the above proof, i.e., matching elements x with congruent elements ax. Let us try to evaluate (mod m) the product $\prod_{x \in R} x$ itself by a similar argument. We know that if $x \in R$ there exists a unique $x' \in R$ such that $xx' \equiv 1 \pmod{m}$. Both the existence and uniqueness of x' are guaranteed by (251), which says that as x' runs through R, xx' runs through a reduced residue system modulo m, and so must hit 1 exactly once. The idea is to pair up the factors of $\prod_{x \in R} x$ in this way, each pair multiplying to 1 (mod m).

We seem to have proved $\prod_{x \in R} x \equiv 1 \pmod{m}$ in the above paragraph, but a little care is needed. We must consider the possibility that x pairs with itself; i.e., $x^2 \equiv 1 \pmod{m}$. This certainly is the case for $x \equiv \pm 1 \pmod{m}$, and perhaps for other x's. For example $4^2 \equiv 1 \pmod{15}$. The situation is simplified when the modulus is prime; we saw earlier that $x^2 \equiv a \pmod{p}$ has at most 2 solutions. A prime modulus p also has the advantage of a very explicit reduced residue system, namely $1, 2, \ldots, p - 1$.

If $p = 11$, for example, we can pair off everything except 1 and 10. To be explicit,

$$2 \cdot 6 \equiv 3 \cdot 4 \equiv 5 \cdot 9 \equiv 7 \cdot 8 \equiv 1 \pmod{11}.$$

Thus $1 \cdot 2 \cdot 3 \cdot 4 \cdot 5 \cdot 6 \cdot 7 \cdot 8 \cdot 9 \cdot 10 = 1(2 \cdot 6)(3 \cdot 4)(5 \cdot 9)(7 \cdot 8)10 \equiv -1$ (mod 11).

(457) **Wilson's Theorem.** If p is prime, then

$$(p - 1)! \equiv -1 \pmod{p}.$$

(458) **Exercise.** Let $m > 1$. Show that $(m - 1)! \equiv -1 \pmod{m}$ if and only if m is prime.

(459) **Exercise.** Illustrate the proof of Wilson's Theorem for $p = 13$ by pairing off the integers $1, 2, \ldots, 12$.

(460) *Note.* The proof given for Wilson's Theorem breaks down for $p = 2$ since then $1 \equiv -1 \pmod{p}$. The result still holds, however. If $p = 3$ there is no pairing to do since $1, -1$ comprises a whole reduced residue system modulo p.

43 A THEOREM ABOUT QUADRATIC RESIDUES

(461) Our proof of Wilson's Theorem was based on the fact that if $p \nmid x$, then $xx' \equiv 1 \pmod{p}$ has a unique solution x' in any reduced residue system modulo

p. The same goes for $xx' \equiv a \pmod{p}$, where $p \nmid a$. Let us see what we can make out of this.

We are here considering $(p - 1)!$ in a new way, pairing elements, the product of which is congruent to $a \pmod p$. Again we must pay special attention to elements x pairing with themselves. Such x satisfy $x^2 \equiv a \pmod p$. Of course we have already seen that a complete solution to this congruence has 2 or 0 elements, depending on whether a is a quadratic residue modulo p or not, so long as p is an *odd* prime.

The situation is simplest when a is not a quadratic residue. Then each of the integers $1, 2, \ldots, p - 1$ pairs with another, the product of the pair being congruent to $a \pmod p$. Since $p - 1$ integers produce $(p - 1)/2$ pairs, we have

$$(p - 1)! \equiv a^{(p-1)/2} \pmod p.$$

If a *is* a quadratic residue $\pmod p$, on the other hand, there exists x_0 such that $x_0^2 = (-x_0)^2 \equiv a \pmod p$. If p is odd, then $x_0 \not\equiv -x_0 \pmod p$. Let us assume p is odd. Deleting x_0 and $-x_0$ from the set $\{1, 2, \ldots, p - 1\}$ leaves $p - 3$ integers to pair up. Thus

$$(p - 1)! \equiv a^{(p-3)/2} x_0 (-x_0) \equiv -x_0^2\, a^{(p-3)/2} \equiv -a \cdot a^{(p-3)/2}$$
$$\equiv -a^{(p-1)/2} \pmod p.$$

Since $(p - 1)! \equiv -1 \pmod p$ by Wilson's Theorem, we have proved

(462) **THEOREM.** Let p be an odd prime and suppose $p \nmid a$. Then $a^{(p-1)/2}$ is congruent to 1 or $-1 \pmod p$ according as a is or is not a quadratic residue modulo p.

(463) **Exercise.** Solve $9^{20} \equiv x \pmod{41}$, $|x| = 1$.

(464) **Exercise.** Determine the quadratic residues a modulo 11, $1 \leqslant a \leqslant 5$, both from the definition and by means of (462).

(465) **Exercise.** Is 2 a quadratic residue $\pmod{19}$?

(466) **Exercise.** Is 3 a quadratic residue $\pmod{13}$?

(467) **Exercise.** Is 10 a quadratic residue $\pmod{17}$?

(468) **Exercise.** Is -1 a quadratic residue $\pmod{67}$?

(469) **Exercise*.** Prove that if p is prime, then -1 is a quadratic residue $\pmod p$ if and only if $p = 2$ or $p \equiv 1 \pmod 4$.

(470) Although in Theorem (462) we have unexpectedly come upon a test for identifying quadratic residues, its practical value is somewhat limited when p is of any size. (An exception is $a = -1$, as the last exercise shows.) Shortcuts are possible, however.

Let us for example test whether 3 is a quadratic residue modulo the prime 29. We must compute 3^{14} (mod 29). We have $3^3 = 27 \equiv -2$ (mod 29). Thus $3^6 \equiv (-2)^2 \equiv 4$ (mod 29) and $3^{12} \equiv 4^2 \equiv 16$ (mod 29). Then $3^{14} \equiv 9 \cdot 16 = 144 \equiv -1$ (mod 29). Thus 3 is not a quadratic residue modulo 29.

(471) **Exercise.** Is 3 a quadratic residue (mod 31)?

(472) **Exercise.** Is 2 a quadratic residue (mod 61)?

(473) **Exercise.** Is 5 a quadratic residue (mod 127)?

(474) **Exercise.** Is 4 a quadratic residue (mod 293)?

44 THE DEGREE OF A CONGRUENCE

(475) If p is an odd prime and $p \nmid a$, it should be no surprise to us that $a^{(p-1)/2} \equiv \pm 1$ (mod p). After all, $(a^{(p-1)/2})^2 = a^{p-1} \equiv 1$ (mod p) by Euler's Theorem. Thus $a^{(p-1)/2}$ is a solution to $x^2 \equiv 1$ (mod p), and we know this congruence has only the solutions $x \equiv \pm 1$ (mod p).

We can look at $a^{(p-1)/2}$ in still another way. If a is a quadratic residue modulo p, then $a \equiv x^2$ (mod p) for some x. Thus $a^{(p-1)/2} \equiv (x^2)^{(p-1)/2} = x^{p-1} \equiv 1$ (mod p), again by Euler's Theorem.

We almost have an alternate proof of Theorem (462). We have shown above that if p is an odd prime and $p \nmid a$, then $a^{(p-1)/2} \equiv \pm 1$ (mod p), with $+1$ in the case that a is a quadratic residue (mod p). The only thing we haven't shown is that $a^{(p-1)/2} \equiv -1$ (mod p) when a is a nonresidue.

Notice that we have found $(p - 1)/2$ solutions to the congruence $x^{(p-1)/2} \equiv 1$ (mod p), namely the $(p - 1)/2$ quadratic residues modulo p. (See (428).) We want to show that a complete solution to $x^{(p-1)/2} \equiv 1$ (mod p) has *at most* $(p - 1)/2$ elements. It would seem natural to call $x^{(p-1)/2} - 1 \equiv 0$ (mod p) a congruence of degree $(p - 1)/2$, since $x^{(p-1)/2} - 1$ is a polynomial of degree $(p - 1)/2$. Then, in analogy with the theorem of algebra that says that a polynomial of degree n has at most n zeros (i.e., numbers which give 0 when substituted into the polynomial), we might hope to prove that a complete solution of a congruence of degree n has at most n elements. This would imply that $a^{(p-1)/2} \equiv 1$ (mod p) only for a equal to one of the $(p - 1)/2$ quadratic residues (mod p), forcing $a^{(p-1)/2} \equiv -1$ (mod p) for nonresidues.

(476) A theorem such as the one suggested above can be proved, but some care is needed, both in finding proper hypotheses and in defining "the degree of a congruence." The natural thing is to define the degree of the congruence $P(x) \equiv 0$ (mod m) to be n in case $P(x)$ is a polynomial of degree n. An example will show that such a definition actually has unnatural consequences. Consider the congruence

$$6x^3 - 2x^2 + x - 1 \equiv 0 \,(\text{mod } 3).$$

By our tentative definition this would be a congruence of degree 3. In any attempt to solve this congruence, however, we would immediately drop the first term, since $6x^3 \equiv 0 \pmod 3$ no matter what x is.

As a more extreme example, consider

$$5x^3 + 10x^2 - 15 \equiv 0 \pmod 5.$$

Our tentative definition would make this a congruence of degree 3. Yet a complete solution clearly has 5 elements. We see our tentative definition is inadequate for proving our tentative theorem that a congruence of degree n has at most n solutions. The following definition gets around the difficulties just mentioned.

(477) **Definition.** Consider the polynomial $P(x) = a_n x^n + a_{n-1} x^{n-1} + \ldots + a_0$. We say the congruence

$$P(x) \equiv 0 \pmod m$$

has *degree* k in case k is the largest subscript such that $m \nmid a_k$. If m divides all the coefficients of $P(x)$ we say the congruence has no degree.

(478) **Examples.** The degree of $3x^2 + 2x + 6 \equiv 0 \pmod 6$ is 2. The degree of $4x^3 + 5x^2 - 2x + 1 \equiv 0 \pmod 2$ is 2. The degree of $10x^4 + 5x^2 - 3x + 1 \equiv 0 \pmod 5$ is 1. The degree of $11x^3 + 1 \equiv 0 \pmod{11}$ is 0. The congruence $8x^2 - 12 \equiv 0 \pmod 4$ has no degree.

(479) **Exercise.** Determine the degrees of the following congruences.
  (a) $12x^3 + 8x^2 + 6x \equiv 0 \pmod 8$
  (b) $12x^3 + 8x^2 + 6x \equiv 0 \pmod 6$
  (c) $12x^3 + 8x^2 + 6x \equiv 0 \pmod 4$
  (d) $12x^3 + 8x^2 + 6x \equiv 0 \pmod 2$
  (e) $5 \equiv 0 \pmod 3$.

45 THE NUMBER OF SOLUTIONS OF A CONGRUENCE OF DEGREE n

(480) Our goal is to prove that a complete solution of a congruence of degree n has at most n elements. A good place to start is $n = 1$. We have a congruence of the form

$$a_1 x + a_0 \equiv 0 \pmod m,$$

where $m \nmid a_1$. (Of course higher powers of x could appear with coefficients divisible by m, but since these in no way affect solutions, we ignore them.) Our congruence can be written as $a_1 x \equiv -a_0 \pmod m$. We have treated such congruences before. Theorem (327) states that a complete solution has (a_1, m) elements or 0 elements depending on whether $(a_1, m) \mid -a_0$ or not. For example $2x - 2 \equiv 0 \pmod 6$ has the complete solution $x = 1, 4$.

This is hardly encouraging for our theorem. We have just written a congruence of degree 1 with 2 incongruent solutions. The trouble comes in our example because we have $(a_1, m) > 1$; otherwise (327) says there is at most 1 solution. We may be tempted to blame our definition of the degree of a congruence. Perhaps we should have made the degree of

$$a_n x^n + \cdots + a_0 \equiv 0 \ (\text{mod } m)$$

to be k if k is the greatest subscript such that $(a_k, m) = 1$.

Changing the definition of the degree of a congruence in this way would have at least two drawbacks. One is that it would leave many nontrivial congruences, such as

$$3x^2 + 2x + 4 \equiv 0 \ (\text{mod } 6),$$

with no degree at all. The second drawback is that the theorem we are after still wouldn't be true.

To see the last assertion we must look at a congruence which has degree 2 by either Definition (477) or our proposed amendment. Consider

$$x^2 - 1 \equiv 0 \ (\text{mod } 15).$$

A complete solution is $x = 1, 4, -1, -4$. *Four* solutions, to a congruence of degree 2 by practically any definition that comes to mind.

(481) Unpromising as the preceding investigations may appear, something can be salvaged. Theorem (416) provides hope. It states that what we now call a congruence of degree 2 has 0, 1, or 2 elements in a complete solution, *if the modulus is an odd prime*. (Of course any congruence modulo 2 has at most 2 solutions.) Likewise, since if p is prime $p{\nmid}a$ implies $(p,a) = 1$, Theorem (327) says a complete solution of any congruence of degree 1 has either 0 or 1 element, *if the modulus is prime*.

We see the truth for $n = 1$ and $n = 2$ of

(482) **Lagrange's Theorem.** If p is prime, then a complete solution of any congruence of degree n modulo p has at most n elements.

46 THE DIVISION ALGORITHM FOR POLYNOMIALS

(483) Note that the preceding theorem is sufficient to provide an alternate proof of (462), which was why we got into this question in the first place.

Since (482) is the analogue of a theorem about solutions to polynomial *equations*, we may hope that the proofs are also analogous, much as our analysis of second degree congruences resembled that of the quadratic equation. Why does a polynomial equation of degree n have at most n roots? We need someone who knows something about the theory of equations.

Expert: That's me. What do you want to know?

Us: Just this. Suppose f is a polynomial of degree n. How do you know it has at most n roots?

Expert: How could it have more? Each root means a linear factor of $f(x)$. More than n roots would mean f had degree more than n.

Us: Would you explain that a little?

Expert: Suppose x_1 is a root of $f(x) = 0$. Then $f(x) = (x - x_1)g(x)$, where $g(x)$ is another polynomial. If x_2 is another root, then plugging it in f shows $g(x_2) = 0$, so $g(x) = (x - x_2)h(x)$ in the same way. Then $f(x) = (x - x_1) \cdot (x - x_2)h(x)$, and the degree of f must be at least 2. Keep it up. If you can take out more than n factors, then the degree of f must also be more than n.

Us: We see. But back up a bit. Why does $f(x_1) = 0$ imply $f(x) = (x - x_1)g(x)$?

Expert: Don't be simple. Say $f(x) = (x - x_1)g(x) + r$, where r is a constant. Plug in x_1 for x. Since $f(x_1) = 0$, you see $r = 0$.

Us: That looks familiar. Something like the Division Algorithm. But how do you know r is a constant?

Expert: The remainder always has degree less than the thing you divide by. Since $x - x_1$ had degree 1, r must be a constant.

Us: You seem to be using some general theorem.

Expert: Sure I am. The one that says if $a(x)$ and $b(x)$ are any two polynomials, then there exist polynomials $q(x)$ and $r(x)$ such that

$$b(x) = q(x)a(x) + r(x),$$

where either $r(x) = 0$ or else the degree of $r(x)$ is less than that of $a(x)$. In any particular case you can use long division. Say $a(x) = x^2 + x - 1$ and $b(x) = x^4 + 2x^3 + 15$, for example. You know how it goes:

$$
\begin{array}{r}
x^2 + x \phantom{{}+ 15} \\
x^2 + x - 1 \overline{\smash{)}\, x^4 + 2x^3 \phantom{{}+x^2{}} + 15} \\
\underline{x^4 + x^3 - x^2 \phantom{{}+ 15}} \\
x^3 + x^2 + 15 \\
\underline{x^3 + x^2 - x } \\
x + 15.
\end{array}
$$

You see $q(x)$ comes out $x^2 + x$ and $r(x)$ is $x + 15$. It always works, although I guess that's not a proof. You could probably prove it by induction.

Us: We'll give it a try. One more thing, though. We're only interested in polynomials with integral coefficients. Is your theorem still good for them?

Expert: Afraid not. Say $a(x) = 2x - 1$ and $b(x) = x^2 + 1$.

$$
\begin{array}{r}
x/2 + 1/4 \\
2x - 1 \overline{\smash{)}\, x^2 + 1 } \\
\underline{x^2 - x/2 } \\
x/2 + 1 \\
\underline{x/2 - 1/4} \\
5/4
\end{array}
$$

Neither the quotient nor remainder are integral polynomials. No way to get around it, either. If $x^2 + 1 = q(x)(2x - 1) + r(x)$ where r is a constant, the leading coefficient of $q(x)$ just has to be $\frac{1}{2}$. By the "leading coefficient" I mean the coefficient of the highest power of x, of course. It's the coefficient 2 in $a(x)$ that gives you the trouble. Divide by a polynomial with leading coefficient 1 and you'll never get any fractions—just think of the long division. Which—come to think of it—is really all you need for that "at most n roots" business, since you're only dividing by things like $x - x_1$.

In light of the above conversation, let us delay our proof of (482) to prove

(484) **THEOREM.** Suppose $a(x)$ and $b(x)$ are integral polynomials and the co-efficient of the highest power of x in $a(x)$ is 1. Then there exist integral polynomials $q(x)$ and $r(x)$, where $r(x)$ is either 0 or has degree less then the degree of $a(x)$, such that

$$b(x) = q(x)a(x) + r(x).$$

Proof. It would be pleasant in what follows if we could manage to deal only with integral polynomials, since we want $q(x)$ and $r(x)$ to be such. If an induction proof is really in order, it must be induction "on" something; i.e., we must decide what n shall be. The most likely candidates are the degrees of $a(x)$ and $b(x)$. Choosing the latter at least allows us to get started. Since always

$$b(x) = 0 \cdot a(x) + b(x),$$

whenever the degree of $b(x)$ is less than that of $a(x)$ taking $q(x) = 0$ and $r(x) = b(x)$ works. In particular, if the degree of $b(x)$ is 0 then the theorem is true. (If the degree of $a(x)$ is also 0 then $a(x)$ must be the constant polynomial 1, and $b(x) = a(x)b(x) + 0$.)

Now let us assume the degree of $b(x)$ is n, which is greater than or equal to the degree of $a(x)$. If we use the form of mathematical induction of (169)(b) we may assume the theorem true for all polynomials $b^*(x)$ having degree less than n. This suggests somehow writing $b(x)$ in terms of such a polynomial of lower degree.

How would we start to use long division to divide $b(x)$ by $a(x)$?

$$a(x) \overline{)b(x)}^{\,(?)}$$

We would compute (?), which is meant to represent the first term in the quotient, by dividing the term of $b(x)$ involving the highest power of x by the corresponding term of $a(x)$. This will yield an integral monomial, since the degree of $a(x)$ is less than that of $b(x)$ and the leading coefficient of $a(x)$ is 1. Let us suppose $b(x) = b_n x^n + b_{n-1} x^{n-1} + \cdots + b_0$ and $a(x) = x^m + a_{m-1} x^{m-1} + \cdots + a_0$, where $m \leqslant n$. Then $(?) = b_n x^{n-m}$.

At the next step of the long division we subtract $b_n x^{n-m} a(x)$ from $b(x)$, and this is the key to our induction proof. Denote by $b^*(x)$ the integral polynomial $b(x) - b_n x^{n-m} a(x)$. Clearly $b^*(x)$ has degree less than n, since we have cancelled the first term of $b(x)$. By our induction hypothesis there exist integral polynomials $q(x)$ and $r(x)$, with the degree of $r(x)$ less than m unless $r(x) = 0$, such that $b^*(x) = q(x)a(x) + r(x)$. Then

$$b(x) = b^*(x) + b_n x^{n-m} a(x) = (q(x) + b_n x^{n-m})a(x) + r(x).$$

This is the required representation of $b(x)$, and our induction proof is completed.

(485) **Exercise.** Find $q(x)$ and $r(x)$ as above for
   (a) $a(x) = x^2 + 3x + 1, b(x) = 2x^2 + 3x + 4$
   (b) $a(x) = x^3 + 3x + 1, b(x) = 2x^2 + 3x + 4$
   (c) $a(x) = x - 1, b(x) = x^4 - 1$.

(486) **Exercise.** Prove or disprove: (484) is still true if the condition "the coefficient of the highest power of x in $a(x)$ is 1" is replaced by "the coefficient of the highest power of x in $a(x)$ divides the coefficient of the highest power of x in $b(x)$."

(487) **Exercise.** Prove that the polynomials $q(x)$ and $r(x)$ guaranteed by (484) are unique.

47 LAGRANGE'S THEOREM PROVED

(488) Finally we are ready to return to proving (482), which says that a congruence of degree n with prime modulus has at most n incongruent solutions. Let us suppose $f(x) \equiv 0 \pmod{p}$ is such a congruence. We may as well assume f is a polynomial of degree n, since dropping terms always divisible by p cannot change the set of solutions.

We have already seen that the theorem is true for $n = 1$ or 2, so a proof by induction seems appropriate. We may assume the theorem true for all congruences of degree less than n, just as in the last proof.

Suppose $f(x_1) \equiv 0 \pmod{p}$. We would like to divide f by $x - x_1$, but things are not quite so simple as in the case of equations. Of course Theorem (484) says we can write $f(x)$ as $(x - x_1)q(x) + r$, where r is a constant and everything is integral. Substituting x_1 for x then gives us

$$f(x_1) = (x_1 - x_1)q(x_1) + r = r \equiv 0 \pmod{p}.$$

Thus we cannot conclude that $r = 0$, but only that $p|r$. This is no great loss, however, since

$$f(x) - r = (x - x_1)q(x) \equiv 0 \pmod{p}$$

clearly has exactly the same solutions as $f(x) \equiv 0 \pmod{p}$.

We could continue to divide out linear factors, as in the Expert's argument in (483), but a slicker way is to use our induction hypothesis. Since the degree of $q(x)$ is $n - 1$ it tells us that $q(x) \equiv 0 \pmod{p}$ has at most $n - 1$ incongruent solutions.

How many solutions can $(x - x_1)q(x) \equiv 0 \pmod{p}$ have? If x_2 is a solution, then $p \mid (x_2 - x_1)q(x_2)$. Since p is prime $p \mid x_2 - x_1$ or $p \mid q(x_2)$. We see $x_2 \equiv x_1 \pmod{p}$ or else x_2 is congruent to one of the $n - 1$ or fewer elements in a complete solution to $q(x) \equiv 0 \pmod{p}$. Thus a complete solution to $f(x) \equiv 0 \pmod{p}$ has at most n elements.

There is a slight gap in the argument given above, although the theorem is true enough. The reader is invited to find it if he hasn't already. It will be revealed in (496).

(489) **Exercise.** Suppose p is prime, $n \geq p$, and $0 \leq k \leq p$. Show there exists a congruence of degree n such that a complete solution contains exactly k elements.

(490) **Exercise.** Show that a complete solution of a polynomial congruence modulo 6 cannot have exactly 5 elements.

(491) **Exercise.** Suppose $m > 1$ has the property that if $0 \leq k \leq m$ then there exists a polynomial congruence having exactly k elements in a complete solution. Show m is prime.

(492) **Exercise.** Let M be any integer. Show there exists a polynomial congruence of degree 2 such that a complete solution has more than M elements.

(493) **Exercise.** How many elements are there in a complete solution to $x(x^2 - 1)(x - 2) \equiv 0 \pmod{120}$?

(494) **Exercise.** Let $\epsilon > 0$. Show that there are positive integers m and n such that $n/m < \epsilon$ and such that there is a congruence of degree n with modulus m satisfied by every integer.

(495) **Exercise.** Suppose p is an odd prime and $0 \leq k \leq 2p$. Show there exists no polynomial congruence modulo $2p$ such that a complete solution has exactly k elements if and only if $k > p$ and k is odd.

(496) The gap in the proof given in (488) came when we concluded that $q(x) \equiv 0 \pmod{p}$ had at most $n - 1$ solutions from the fact that $q(x)$ was a polynomial of degree $n - 1$. Recall that the degree of the congruence $q(x) \equiv 0 \pmod{p}$ is not necessarily the same as the degree of $q(x)$ as a polynomial. Sure, you say, but so what? The degree of the congruence certainly cannot be *greater* than $n - 1$, and if it is less, then the induction hypothesis tells us all the more that it has at most $n - 1$ solutions.

True enough, I retort, unless $q(x) \equiv 0 \pmod{p}$ has no degree at all. A sobering thought, but easily dealt with. To say $q(x) \equiv 0 \pmod{p}$ has no degree means that all the coefficients of $q(x)$ are divisible by p. But then the same is true of

$$f(x) = (x - x_1)q(x) + r,$$

since $p \mid r$. Then $f(x) \equiv 0 \pmod{p}$ has no degree, contrary to the hypothesis that it is a congruence of degree n. It cannot happen.

(497) **Exercise.** Write out the alternate proof of Theorem (462), assuming (482).

(498) **Exercise.** Write out an alternate proof of Theorem (449), based on Theorem (484). [*Hint:* Divide $P(x)$ by $x^{p-1} - 1$.]

48 COUNTING INTEGERS OF A GIVEN ORDER

(499) Knowing that a congruence of degree n modulo p can have at most n incongruent solutions gives us information about the number of elements in a reduced residue system modulo p having various orders. For example the congruence $x^2 \equiv 1 \pmod{p}$ has degree 2 and therefore has at most two solutions among $1, 2, \ldots, p - 1$. Thus at most two of these integers have order 2 modulo p. In fact $x = 1$ provides one of the solutions, and 1 has order 1 modulo p; this leaves at most one element of order 2. Of course if $p > 2$ then -1 has order 2 modulo p. We see that *if p is an odd prime then any reduced residue system modulo p contains exactly one element of order 2.*

Counting elements of order 3 is a little less certain. We know $x^3 \equiv 1 \pmod{p}$ has at most three solutions, and one of these is $x = 1$. Thus *there are at most two elements of order 3 in any reduced residue system modulo the prime p.* Of course we cannot say for sure that there will be *any* elements or order 3. Since we proved earlier that the order of an integer modulo m always divides $\varphi(m)$ (see (440) and what follows), there are certainly no elements of order 3 unless $3 \mid \varphi(p) = p - 1$. No integer can have order 3 modulo 5, for example.

(500) **Exercise.** Show that if p is prime then the number of integers among $1, 2, \ldots, p - 1$ having order 3 modulo p is either 0 or 2. [*Hint:* Show that if a has order 3 then so does a^2.]

(501) Let us see how far we can carry arguments such as the above with a definite example, say $p = 7$. We know that if a has order g modulo 7, then $g \mid \varphi(7) = 6$. Thus the only possible orders are 1, 2, 3, and 6. From our investigations above we can partially fill out a little table;

g	Number of elements of order g in a reduced residue system
1	1
2	1
3	at most 2
6	?

What can we say about the number of elements of order 6? Each satisfies $x^6 \equiv 1 \pmod{7}$; thus there are at most 6 of them. Not all solutions of this congruence have order 6, however; the solutions $x = 1$ and $x = -1$ do not, for example. This leaves at most 4 elements of order 6.

Notice that any element of order 3 also is a solution, since $3 \mid 6$. [See

Proposition (437).] If there are two elements of order 3, this leaves at most 2 elements of order 6.

Of course we don't know if there are *any* elements of order 3. The existence of elements of order 6 implies elements of order 3, however. For if a has order 6, then a^2 has order 3, since $(a^2)^3 \equiv 1$ while $(a^2)^2 \not\equiv 1 \pmod 7$. Likewise a^4 has order 3.

Since there are either 0 or 2 elements of order 6, our table becomes

g	Number of elements of order g
1	1
2	1
3	at most 2
6	at most 2

Of course each of the 6 elements of a reduced residue system modulo 6 has *some* order; each must be counted somewhere in the above table. But the only way the right-hand column can add up to 6 is if there are *exactly* 2 elements of order 3 and *exactly* 2 elements of order 6. This must be the case.

We have figured out the number of elements of each order in a reduced residue system modulo 7 without once resorting to the dirty business of raising numbers to powers. Isn't mathematics wonderful? Back in (429) we did this the hard way (?) and found that 1 has order 1, 6 has order 2, 2 and 4 have order 3, and 3 and 5 have order 6 modulo 7.

(502) **Exercise.** Make arguments similar to the above to determine the number of elements of each order in a reduced residue system modulo 5. Compare with the result of direct calculation.

(503) **Exercise.** Determine the number of elements of each order in a reduced residue system modulo 11. Compare with the results of Exercise (434).

(504) **Exercise.** Show that if p and q are primes and $q \mid p - 1$, then any reduced residue system modulo p contains either 0 or $q - 1$ elements of order q modulo p.

49 THE COUNT IN GENERAL

(505) It appears from (501) that we may be able to say something in general about the number of elements of order g in a reduced residue system modulo p, where p is prime. To save words, let us denote this number by $N(g)$. We have seen that $N(g) = 0$ unless $g \mid \varphi(p) = p - 1$, which we therefore assume.

As before, we note that any element of order g is a solution to

(*) $$x^g \equiv 1 \pmod p.$$

Thus Theorem (482) tells us that

$$N(g) \leqslant g.$$

Let us suppose there exists an element a of order g modulo p. Then any power of a also satisfies (*), since

$$(a^k)^g = (a^g)^k \equiv 1^k \equiv 1 \ (\text{mod } p).$$

But Proposition (437) says that the numbers $a, a^2, \ldots, a^g$ are all incongruent modulo p. Since these integers comprise g incongruent solutions to (*), a congruence of degree g, they must be a complete solution.

We see that to count $N(g)$ it suffices to examine the orders of the elements $a, a^2, \ldots, a^g$, casting out those with orders less than g. If $k > 1$ is a divisor of g, for example, then a^k does not have order g, since

$$(a^k)^{g/k} = a^g \equiv 1 \ (\text{mod } p)$$

and $g/k < g$. Recalling that $\tau(g)$ represents the number of positive divisors of g (including the divisor 1), we see that

$$N(g) \leqslant g - (\tau(g) - 1).$$

Even more of the integers $a, a^2, \ldots, a^g$ can be eliminated. Suppose $d > 1$ is a common divisor of k and g. Then

$$(a^k)^{g/d} = (a^g)^{k/d} \equiv 1^{k/d} \equiv 1 \ (\text{mod } p),$$

where g/d is an integer less than g. Thus the order of a^k is less than g. We see that a^k cannot have order g unless $(k,g) = 1$. Of course the number of such k, $1 \leqslant k \leqslant g$, is just $\varphi(g)$. We have proved

$$N(g) \leqslant \varphi(g).$$

(506) **Exercise.** Prove that, in the above notation, $N(g) = 0$ or $N(g) = \varphi(g)$.

(507) Each of the $p - 1$ elements in a reduced residue system modulo p has some order g, and we have seen that $N(g) = 0$ if $g \nmid p - 1$ and $N(g) \leqslant \varphi(g)$ if $g | p - 1$. Thus, summing over all possible orders, we have

$$p - 1 = \Sigma_{g|p-1} N(g) \leqslant \Sigma_{g|p-1} \varphi(g).$$

(508) The reader is no stranger to sums such as the above. In fact, Exercise (223) required the evaluation of exactly the sum $\Sigma_{d|n} \varphi(d)$, which appears on the right. Nevertheless, we will go through that evaluation here, as a reminder of the technique we developed.

First we note that since φ is a multiplicative function (a fact we've proved many times), so is the function F defined by

$$F(n) = \Sigma_{d|n} \varphi(d).$$

We proceed to evaluate F at prime powers. Recall that if q is prime, then $\varphi(q^k) = (q - 1)q^{k-1}$. Thus

$$F(q^k) = \Sigma_{d|q^k}\ \varphi(d) = \varphi(1) + \varphi(q) + \varphi(q^2) + \cdots + \varphi(q^k)$$
$$= 1 + (q-1) + (q-1)q + \cdots + (q-1)q^{k-1}$$
$$= 1 + (q-1) \cdot \frac{q^k - 1}{q - 1} = q^k,$$

where we used (125) to sum the geometric series $(q-1) + (q-1)q + \cdots + (q-1)q^{k-1}$.

Now suppose $n = q_1^{k_1} q_2^{k_2} \ldots q_t^{k_t}$, the q's distinct primes. Since F is multiplicative we have

$$F(n) = F(q_1^{k_1})F(q_2^{k_2})\ldots F(q_t^{k_t}) = q_1^{k_1} q_2^{k_2} \ldots q_t^{k_t} = n.$$

(509) Applying this result to the inequality at the end of (507), we get

$$p - 1 = \Sigma_{g|p-1}\ N(g) \leqslant \Sigma_{g|p-1}\ \varphi(g) = p - 1.$$

Clearly the two sums must be equal. But if we had $N(g) < \varphi(g)$ for any g dividing $p-1$ the first sum would be smaller. We conclude that

(510) **THEOREM.** Suppose p is prime and $g|p-1$. Then there are exactly $\varphi(g)$ elements of order g in any reduced residue system modulo p.

(511) **Exercise.** Show that it is *not* true that if $g|\varphi(m)$, then there are exactly $\varphi(g)$ elements of order g in any reduced residue system modulo m. Where does the proof given above go wrong for non-prime moduli?

(512) **Exercise.** Determine the number of elements of each order in a reduced residue system modulo 13.

(513) **Exercise.** Determine the number of elements of each order in a reduced residue system modulo 17.

(514) **Exercise.** Determine the number of elements of each order in a reduced residue system modulo 18.

(515) **Exercise.** Determine the number of elements of each order in a reduced residue system modulo 20.

(516) **Exercise*.** Suppose m and k are positive integers, and a has order g modulo m. Show that if $(k,g) = 1$ then a^k also has order g modulo m. [*Hint*: If $(a^k)^h \equiv 1 \pmod{m}$, then $g|kh$ by (437). Use (249).]

(517) **Exercise.** Suppose a has order g modulo m, and k is a positive integer relatively prime to $\varphi(m)$. Show that a^k has order g also.

chapter 6

The Theory of Primitive Roots

Certain moduli have reduced residue systems consisting of powers of a fixed integer. We will see how this situation can be advantageous, and determine for which moduli it occurs.

50 PRIMITIVE ROOTS

(518) An important consequence of Theorem (510) is that if p is prime there exist integers A of the maximum possible order modulo p, namely $\varphi(p) = p - 1$. Thus $A, A^2, \ldots, A^{p-1}$ constitute $p - 1$ incongruent integers modulo p, and so are a reduced residue system.

The existence of such an explicit reduced residue system is a happy state of affairs whether the modulus m is prime or not. It occurs whenever there is an integer A of order $\varphi(m)$ modulo m, for then $A, A^2, \ldots, A^{\varphi(m)}$ form a reduced residue system modulo m by (437). If $g|\varphi(m)$, then the order of $A^{\varphi(m)/g}$ is easily seen to be g, since the smallest positive integer k such that $\varphi(m)| k\varphi(m)/g$ is clearly g. We see that the existence of an integer of order $\varphi(m)$ implies that of integers of all orders dividing $\varphi(m)$.

Even more is true. Exercise (516) says that if a has order g modulo m, then so does a^k so long as $(k,g) = 1$. Since the integers $A^{k\varphi(m)/g}$ are incongruent for $1 \leqslant k \leqslant g$, and have order g for $(k,g) = 1$, we see that if $g|\varphi(m)$, then

$$N(g) \geqslant \varphi(g),$$

in the notation of (505). Summing over the divisors of $\varphi(m)$, we get

$$\varphi(m) = \Sigma_{g|\varphi(m)} \, N(g) \geqslant \Sigma_{g|\varphi(m)} \, \varphi(g).$$

Since the summation on the right equals $\varphi(m)$ by (509) we again conclude that $N(g) = \varphi(g)$ for each g dividing $\varphi(m)$.

(519) **Definition.** We say A is a *primitive root modulo m* in case the order of A modulo m is $\varphi(m)$.

(520) **THEOREM.** Suppose there exists a primitive root modulo m. Then if $g \mid \varphi(m)$, there exist exactly $\varphi(g)$ elements of order g in any reduced residue system modulo m.

(521) **Exercise.** What, if any, are the primitive roots A modulo m, $1 \leqslant A \leqslant m$, for $m = 10, 11, 12$, and 13?

(522) **Exercise.** Find all the primitive roots A modulo 22, $1 \leqslant A \leqslant 22$, given that $A = 7$ is one. [*Hint*: Use (516).]

(523) **Exercise.** Suppose m and k are positive integers, and there exists a primitive root modulo m. Show that a complete solution of $x^k \equiv 1 \pmod{m}$ has $(k, \varphi(m))$ elements. [*Hint*: Let A be a primitive root. If suffices to count the integers t, $1 \leqslant t \leqslant \varphi(m)$, such that A^t satisfies the congruence. This happens if and only if $kt \equiv 0 \pmod{\varphi(m)}$. Use (327).]

51 PRIMITIVE ROOTS MODULO TWICE AN ODD NUMBER

(524) **Proposition.** If p is prime then there exists a primitive root modulo p.

(525) We have already noted that the above Proposition is a consequence of Theorem (510). If *only* primes had primitive roots Theorem (520) would say no more than (510), and its proof would have been a waste of our time. One need not search far to find that such is not the case. The integer 3 is a primitive root modulo 4, since its order is $2 = \varphi(4)$. Likewise 5 has order $2 = \varphi(6)$ modulo 6, and so is a primitive root. Of course generating a reduced residue system with only 2 elements is no great thing. Finding a primitive root modulo 6 hardly seems more difficult than finding one modulo 3, since in both cases only an element of order 2 is needed. Since $\varphi(2m) = \varphi(m)$ whenever m is odd, perhaps in such cases a primitive root modulo m will also serve for $2m$.

Consider $m = 7$, for example. Note that $3^1 \equiv 3$, $3^2 \equiv 2$, $3^3 \equiv 6$, $3^4 \equiv 4$, $3^5 \equiv 5$, and $3^6 \equiv 1 \pmod 7$; thus 3 is a primitive root. Since the integers 3^k, $1 \leqslant k \leqslant 6$ are incongruent modulo 7, they are certainly incongruent modulo 14. But $\varphi(14) = 6$ also. We see 3 is a primitive root modulo 14.

Now suppose $m = 11$. Here as $k = 1, 2, \ldots, 10$ we find $2^k \equiv 2, 4, 8, 5, 10, 9, 7, 3, 6, 1 \pmod{11}$; thus 2 is a primitive root. Although the first 10 powers 2 are incongruent modulo 22, however, they are *not* a reduced residue system. The trouble is that none are prime to 22. This is easily fixed up by starting with the odd integer $13 = 2 + 11$ instead of 2. Since $13 \equiv 2 \pmod{11}$, 13 is also a primitive root modulo 11. The first 10 powers of 13, being incongruent *and* relatively prime to 22, form a reduced residue system modulo 22.

(526) **Proposition.** If the odd integer m has a primitive root, then so does $2m$.

Proof. Suppose A is a primitive root modulo m. Let $B = A$ or $A + m$, whichever is odd. The integers B^k, $1 \leqslant k \leqslant \varphi(m)$, are incongruent modulo $2m$ because they are incongruent modulo m. Since $(B, m) = 1$ and $(B, 2) = 1$, $(B, 2m) = 1$; and the same goes for all powers of B. But $\varphi(2m) = \varphi(m)$; thus B generates a reduced residue system modulo $2m$. We see B is a primitive root modulo $2m$.

(527) **Examples.** Let us find all the primitive roots modulo 26. We know such exist since 13 is prime. In fact, we know there are $\varphi(\varphi(26)) = 4$ of them.

First we look for a primitive root modulo 13. The integer 2 is easiest to test. We have $2^2 \equiv 4$, $2^3 \equiv 8$, $2^4 \equiv 16 \equiv 3$, $2^5 \equiv 6$, and $2^6 \equiv 12$ (mod 13). Since the order of any integer must divide $\varphi(13) = 12$, there is no reason to go further. We see 2 does not have order 2, 3, 4, or 6; it must have order 12.

Since each primitive root modulo 13 determines a primitive root modulo 26, as in the proof of (526), we will find all the primitive roots modulo 13 first. We use (516), which says that the primitive roots modulo 13 are the integers 2^k with $1 \leqslant k \leqslant 12$ and $(k, 12) = 1$. Thus the others are $2^5 \equiv 32 \equiv 6$, $2^7 \equiv 2^2 6 \equiv 11$, and $2^{11} \equiv 2^4 11 \equiv 7$ (mod 13).

Finally, the 4 primitive roots modulo 26 are $15 = 2 + 13$, $19 = 6 + 13$, 11, and 7.

As another example let us find a primitive root modulo 59, which is prime. Since $\varphi(59) = 58 = 2 \cdot 29$, any integer prime to 59 has order 1, 2, 29, or 58. Let us try 2 as a primitive root. We note that $2^6 \equiv 64 \equiv 5$ (mod 59). Successively squaring both sides, we get $2^{12} \equiv 25$ and $2^{24} \equiv 625 \equiv 35$ (mod 59). Thus

$$2^{29} \equiv 2^5 \cdot 35 = 1120 \equiv -1 \text{ (mod 59)}.$$

We see 2 is a primitive root. There are 27 more.

(528) **Exercise.** Find all the primitive roots A modulo 34, $1 \leqslant A \leqslant 34$.

(529) **Exercise.** Find all the primitive roots A modulo 25, $1 \leqslant A \leqslant 25$, if any.

(530) **Exercise.** Find all the primitive roots A modulo 21, $1 \leqslant A \leqslant 21$, if any.

(531) **Exercise.** Find the smallest positive primitive root modulo 125, if any.

(532) **Exercise.** Is 2 a primitive root modulo 47? Is 3?

(533) **Exercise.** Is 10 a primitive root modulo 19?

(534) **Exercise.** Prove that if $p \equiv 1$ (mod 4), p prime, and if A is a primitive root modulo p, then so is $-A$.

(535) **Exercise.** Prove that if A is a primitive root modulo m, with $m > 2$, then it is not a quadratic residue modulo m.

(536) **Exercise.** On the basis of what has been proved so far, which of the following moduli definitely have primitive roots: 60, 61, 62, 63, 64, 65?

52 MODULI WITHOUT PRIMITIVE ROOTS

(537) The modulus 2 has a primitive root, and if m does, then so does $2m$. We might hope that, in general, if m_1 and m_2 have primitive roots, then so does $m_1 m_2$.

For A to be a primitive root modulo $m_1 m_2$ we must have that $g = \varphi(m_1 m_2)$ is the smallest solution to

(I) $$A^g \equiv 1 \pmod{m_1 m_2}.$$

This congruence implies the system

(II)
$$A^g \equiv 1 \pmod{m_1}$$
$$A^g \equiv 1 \pmod{m_2}.$$

If $(m_1, m_2) = 1$, in fact, (I) and (II) are equivalent, according to Proposition (261). Let us assume $(m_1, m_2) = 1$.

At least the converse of what we are trying to prove is true; i.e., if A is a primitive root modulo $m_1 m_2$, then A is a primitive root modulo both m_1 and m_2. This is because for the powers of A to run through a reduced residue system modulo $m_1 m_2$ they must enter all congruence classes modulo m_1 prime to m_1; likewise for m_2. To make sure this is clear in the reader's mind, let it be an exercise.

(538) **Exercise***. If $(m_1, m_2) = 1$ and the set S contains a reduced residue system modulo $m_1 m_2$, then S contains a reduced residue system modulo m_1. [*Hint*: Suppose $(j, m_1) = 1$. Show that in fact $s \equiv j \pmod{m_1}$, $s \equiv 1 \pmod{m_2}$ is solvable in S.]

(539) We see that if A is a primitive root modulo $m_1 m_2$, then the order of A modulo m_1 is $\varphi(m_1)$ and the order of A modulo m_2 is $\varphi(m_2)$. Notice that (II) is satisfied by g if and only if g is a common multiple of $\varphi(m_1)$ and $\varphi(m_2)$. The assumption that A is a primitive root modulo $m_1 m_2$ leads us to the conclusion that $\varphi(m_1 m_2)$ is the *least* common multiple of $\varphi(m_1)$ and $\varphi(m_2)$, since (I) and (II) are equivalent. As $\varphi(m_1 m_2) = \varphi(m_1)\varphi(m_2)$, we have proved that

$$\varphi(m_1)\varphi(m_2) = [\varphi(m_1), \varphi(m_2)].$$

We now have the pleasure of invoking the very first theorem of this book, which says that if a and b are positive integers, then

$$[a,b] = ab/(a,b).$$

Such were the simple things we worried about back then. We see $[a,b] = ab$ only if $(a, b) = 1$. Applied to the present matter, this means $\varphi(m_1)$ and $\varphi(m_2)$ are relatively prime.

But this is very unlikely. If the odd prime p divides m_1, for example, then $p - 1$ divides $\varphi(m_1)$; in particular $\varphi(m_1)$ is even. Thus $\varphi(m_1)$ is even unless m_1

is a power of 2. If $k > 1$, however, $\varphi(2^k)$ is also even. We see that $\varphi(m_1)$ is even unless $m_1 = 1$ or 2.

Of course if $\varphi(m_1)$ and $\varphi(m_2)$ are both even, then $(\varphi(m_1), \varphi(m_2)) > 1$ and $m_1 m_2$ cannot have a primitive root. In other words, *if m can be written as $m_1 m_2$, with $(m_1, m_2) = 1$ and both m_1 and m_2 greater than 2, then m does not have a primitive root.*

Let us determine more explicitly which moduli this rules out. Suppose the odd prime p divides m. We can write $m = p^\alpha m'$, where $p \nmid m'$. If $m' > 2$, then m has no primitive root by the above. Thus m' must be 1 or 2, and m is p^α or $2p^\alpha$.

The case remains when m is a power of 2. Then what we have proved above tells us nothing, since if $m = m_1 m_2$ with $(m_1, m_2) = 1$, then $m_1 = 1$ or $m_2 = 1$.

(540) **Proposition.** If the odd prime p divides m, then m has no primitive root unless m is a power of p or twice a power of p.

(541) **Exercise.** On the basis of the propositions we have proved so far, which integers k, $94 \leqslant k \leqslant 103$, definitely do have primitive roots? Which definitely do not?

(542) **Exercise.** Prove or disprove: (538) is true even if the hypothesis $(m_1, m_2) = 1$ is dropped.

(543) **True-False.** Suppose A and B are primitive roots modulo m.

    (a) AB is a primitive root modulo m
    (b) $-A$ is a primitive root modulo m
    (c) $A + B$ is a primitive root modulo m
    (d) $A + B$ is not a primitive root modulo m
    (e) A^2 is not a primitive root modulo m
    (f) If $B \equiv A^k \pmod{m}$, then $(m,k) = 1$
    (g) If $m > 2$, then $A^{\varphi(m)/2} \equiv -1 \pmod{m}$
    (h) If C is a solution to $AC \equiv 1 \pmod{m}$, then C is a primitive root modulo m
    (i) If $C^{\varphi(m)/2} \equiv -1 \pmod{m}$, then C is a primitive root modulo m.

53 PRIMITIVE ROOTS MODULO POWERS OF 2

(544) Since our recent discussion has left out the case of moduli which are powers of 2, let us turn to it now. We have already seen that 1 is a primitive root modulo 2 and 3 is a primitive root modulo 4.

Does 8 have a primitive root? Note that $\varphi(8) = 4$. But $3^2 = 9$, $5^2 = 25$, and $7^2 = 49$ are all congruent to 1 modulo 8. Thus 3, 5, and 7 all have order 2. No integer has order 4. There is no primitive root modulo 8.

This peculiarity of the modulus 8 also wipes out the higher powers of 2.

Suppose A is a primitive root modulo 16, for example. Then various powers of A must be congruent to 3, 5, and 7 modulo 16. These same powers are then also congruent to 3, 5, and 7 modulo 8, which we know is impossible.

The same argument works for higher powers of 2. We leave it to the reader to write out the proof.

(545) **Proposition.** The only powers of 2 having primitive roots are 2 and 4.

(546) There are still moduli to which none of the propositions we have proved apply. One of them is 9. Note that $\varphi(9) = 6$. Also $2^2 \equiv 4$ and $2^3 \equiv 8 \pmod 9$. Since 2 does not have orders 2 or 3 modulo 9, it must be a primitive root.

This calculation enables us to complete the following table of moduli up to 10:

m	Primitive Root?	Reason
2	Yes	Proposition (524)
3	Yes	Proposition (524)
4	Yes	Proposition (545)
5	Yes	Proposition (524)
6	Yes	Propositions (524) and (526)
7	Yes	Proposition (524)
8	No	Proposition (545)
9	Yes	Direct calculation
10	Yes	Propositions (524) and (526).

(547) **Exercise*.** Continue the above table up to $m = 25$. (If previous exercises are used, no calculation should be necessary.)

54 PRIMITIVE ROOTS MODULO p^2

(548) Precisely which moduli m are still in doubt? If two different odd primes divide m then there is no primitive root by Proposition (540). Thus we are left with moduli of the form $2^\alpha p^\beta$, with p an odd prime. Proposition (545) takes care of powers of 2, so we may assume $\beta > 0$. Proposition (540) outlaws $\alpha \geqslant 2$, so we have p^β or $2p^\beta$. Finally, the case $\beta = 1$ is taken care of by Propositions (524) and (526).

The integers we still have to deal with are exactly those of the form p^β and $2p^\beta$, where p is an odd prime and $\beta > 1$. In a few such cases we have looked for primitive roots by direct calculation, namely for $m = 9$ [in (546)], $m = 18$ [Exercise (514)], $m = 25$ [Exercise (529)], and $m = 125$ [Exercise (531)]. In each case a primitive root has been found. Since if p^β has a primitive root, then so does $2p^\beta$, it makes sense to start with powers of odd primes.

(549) **Exercise.** Find the least positive primitive root A modulo m, if any, for $m = 27, m = 49$.

(550) Since in all our examples when m has been a power of an odd prime p we have been able to find a primitive root modulo m, it is a reasonable conjecture

that this is always the case. We know p itself has a primitive root, so this seems a good place to start. Perhaps any primitive root modulo p is also a primitive root modulo powers of p.

At least the converse is true. For suppose A is a primitive root modulo p^β, with $\beta > 1$. Then there must be powers of A congruent to $1, 2, \ldots, p-1$ modulo p^β. These same powers are then congruent to $1, 2, \ldots, p-1$ modulo p. We see A is a primitive root modulo p. This is nearly the same argument used to prove Proposition (545).

The reader may wonder why we bother with a proof such as the above, which, as we have admitted, yields not the proposition we wish to show, but rather its converse. The answer is that *it tells us where to look* for a primitive root modulo p^β, namely, among the primitive roots modulo p. No other candidates need apply. Thus, although it is of no direct help toward the result we seek, our observation may help us find a proof of that result.

As we frequently do, let us simplify the problem as much as possible. Let us start by merely trying to find a primitive root modulo p^2. Suppose A is a primitive root modulo p. Then $g = p - 1$ is the smallest solution to

$$A^g \equiv 1 \ (\text{mod } p).$$

We desire that the smallest solution g to

$$A^g \equiv 1 \ (\text{mod } p^2)$$

should be

$$\varphi(p^2) = p(p - 1).$$

Let us denote by G the order of A modulo p^2. Of course $G | p(p-1)$. On the other hand, $A^G \equiv 1 \ (\text{mod } p^2)$ implies $A^G \equiv 1 \ (\text{mod } p)$, so we see $p - 1 | G$. Since $G/(p-1)$ divides $p(p-1)/(p-1) = p$, $G/(p-1)$ can only be p or 1, and G is either $p(p-1)$ or $p-1$.

Of course $G = p(p-1)$ is exactly what we want; it says A is a primitive root modulo p^2. We somehow have to eliminate the case $G = p-1$. We must show

$$A^{p-1} \equiv 1 \ (\text{mod } p^2)$$

cannot happen.

(551) After a respectable amount of doodling and staring at the last congruence, one may get the idea that we are perhaps trying to prove too much. Maybe not *every* primitive root modulo p is also a primitive root modulo p^2. We used primitive roots modulo m to get primitive roots modulo $2m$ when m was odd in proving (526), but not every primitive root modulo m worked. Perhaps the same is true here.

Why not *count* the number of primitive roots modulo p^2, using Theorem (520)? According to it (and assuming that p^2 does have a primitive root—we're still just groping for a proof, after all) the number of primitive roots modulo p^2 is exactly $\varphi(\varphi(p^2)) = \varphi(p(p-1))$. Since $(p, p-1) = 1$, this number is $\varphi(p) \cdot \varphi(p-1) = (p-1)\varphi(p-1)$.

How many primitive roots are there modulo p? That's easy: $\varphi(\varphi(p)) = \varphi(p - 1)$. But wait a minute—how are we counting things? The number $\varphi(p - 1)$ represents the number of primitive roots in a reduced residue system modulo p. The number $(p - 1)\varphi(p - 1)$ represents the number of primitive roots modulo p^2 in a reduced residue system modulo p^2. This is like comparing the number of Chevrolets in Nebraska to the number of Fords in the United States. To make things fair we should count primitive roots in the same set, say a reduced residue system modulo p^2.

Each congruence class modulo p contains p congruence classes modulo p^2, since the numbers

$$k, k + p, \ldots, k + (p - 1)p$$

are incongruent modulo p^2. Thus the number of primitive roots modulo p in a reduced residue system modulo p^2 is p times $\varphi(p - 1)$.

Summary: Any reduced residue system modulo p^2 contains $p\varphi(p - 1)$ primitive roots modulo p and $(p - 1)\varphi(p - 1)$ primitive roots modulo p^2, assuming p^2 has a primitive root at all.

(552) **Example.** We will illustrate the above by finding all the primitive roots modulo 5 among a reduced residue system modulo 25. The prime 5 has $\varphi(\varphi(5)) = \varphi(4) = 2$ primitive roots in any reduced residue system; the smallest positive primitive roots are 2 and 3. Thus the integers k satisfying $1 \leqslant k \leqslant 25$ which are primitive roots modulo 5 are just

$$2 \quad 7 \quad 12 \quad 17 \quad 22$$
$$3 \quad 8 \quad 13 \quad 18 \quad 23$$

There are a total of $5 \, \varphi(5 - 1) = 5 \cdot 2 = 10$ of them.

Since 25 has a primitive root, exactly $(5 - 1)\varphi(5 - 1) = 4 \cdot 2 = 8$ of the above integers should be primitive roots modulo 25. In Exercise (529) these were found to be the numbers circled below:

② 7 ⑫ ⑰ ㉒
③ ⑧ ⑬ 18 ㉓

(553) **Exercise.** Make a table such as the above showing the primitive roots modulo 7 and modulo 49 among the integers between 1 and 49.

(554) Since $p\varphi(p - 1)$ is always bigger than $(p - 1)\varphi(p - 1)$, there will always be primitive roots modulo p which are not primitive roots modulo p^2. There will be $p\varphi(p - 1) - (p - 1)\varphi(p - 1) = \varphi(p - 1)$ of them, in fact.

Suppose A is a primitive root modulo p. Then there are p integers congruent to A modulo p in any reduced residue system modulo p^2. Since $p > \varphi(p - 1)$, not all of these can belong to the class of integers which are primitive roots modulo p but not modulo p^2. In other words, there must exist an A', $A' \equiv A \pmod{p}$, such that A' is a primitive root modulo p^2. We are keeping in mind, of course, that this is all under the assumption that p^2 has a primitive root, which is really what we are out to prove.

(555) Let us now return to trying to show that p^2 has a primitive root. We suppose A is a primitive root modulo p. We found in (550) that the order of A modulo p^2 is either $p - 1$ or $p(p - 1)$. Since if $p(p - 1)$ is correct then A is a primitive root modulo p^2, we will assume the order of A modulo p^2 is $p - 1$.

   In light of (554) we will search for a primitive root modulo p^2 among the integers $A + kp$, $k = 0, 1, \ldots, p - 1$. It suffices to find that one of these does *not* have order $p - 1$, since the same argument already given shows each has order either $p - 1$ or $p(p - 1)$. We would like a solution k to

(#) $(A + kp)^{p-1} \not\equiv 1 \pmod{p^2}$.

   Our first thought is to use the Binomial Theorem to multiply out the left side of the above congruence. Actually all we need is the weak form of the Binomial Theorem we proved as Proposition (356), which says that

$$(x + ym)^n \equiv x^n + nx^{n-1}ym \pmod{m^2}.$$

Applying this to our situation, we get

$$(A + kp)^{p-1} \equiv A^{p-1} + (p - 1)A^{p-2}kp \equiv A^{p-1} - A^{p-2}kp \pmod{p^2}.$$

Since we are assuming $A^{p-1} \equiv 1 \pmod{p^2}$, we have

$$(A + kp)^{p-1} \equiv 1 - A^{p-2}kp \pmod{p^2}.$$

Thus (#) becomes

$$1 - A^{p-2}kp \not\equiv 1 \pmod{p^2},$$

or

$$- A^{p-2}kp \not\equiv 0 \pmod{p^2}.$$

But since $(A,p) = 1$, this is true for any k except 0; $k = 1$ works, for example. We have proved

(556) **Proposition.** If p is prime then p^2 has a primitive root. In fact, if A is a primitive root modulo p, then either A or $A + p$ is a primitive root modulo p^2.

55 PRIMITIVE ROOTS MODULO p^β

(557) Before trying to prove the existence of primitive roots modulo p^β for $\beta > 2$, let us see what to expect by making a count, such as proved useful previously. Now that we have shown p^2 has a primitive root, we know it has $(p - 1)\varphi(p - 1)$ in any reduced residue system modulo p^2. Since each congruence class modulo p^2 contains p congruence classes modulo p^3, there are $p(p - 1)\varphi(p - 1)$ primitive roots modulo p^2 in any reduced residue system modulo p^3.

Let us assume for the moment that p^3 has a primitive root. Then, according to Theorem (520), there are exactly $\varphi(\varphi(p^3))$ primitive roots in any reduced residue system modulo p^3. But $\varphi(\varphi(p^3)) = \varphi(p^2(p-1)) = \varphi(p^2) \cdot \varphi(p-1) = p(p-1)\varphi(p-1)$. This is the same as the number of primitive roots modulo p^2 in any reduced residue system modulo p^3. Since it is easily seen that any primitive root modulo p^3 is also a primitive root modulo p^2, we have proved that *if there exists any primitive root modulo p^3, then every primitive root modulo p^2 is one.*

(558) **Exercise.** Prove that if $\beta > 2$ and if there exists a primitive root modulo p^β, p prime, then every primitive root modulo $p^{\beta-1}$ is a primitive root modulo p^β.

(559) We have proved that p^β has a primitive root for $\beta = 1$ and $\beta = 2$. The time seems ripe for an induction proof. The last exercise shows that the primitive roots of p^β, if any, are just the primitive roots of $p^{\beta-1}$, assuming $\beta > 2$. This will be our starting point.

Let us assume that p^n has the primitive root A, and try to show that A is also a primitive root modulo p^{n+1}. Let G be the order of A modulo p^{n+1}. We know $G \mid \varphi(p^{n+1}) = p^n(p-1)$. Since $A^G \equiv 1 \pmod{p^{n+1}}$ implies $A^G \equiv 1 \pmod{p^n}$, and since A has order $p^{n-1}(p-1)$ modulo p^n, we see $p^{n-1}(p-1) \mid G$. Thus $G = p^{n-1}(p-1)$ or $G = p^n(p-1)$. Since the latter says A is a primitive root modulo p^{n+1}, we will assume the former and look for a contradiction.

The fact that A is a primitive root modulo p^n tells us something about the form of $A^{p^{n-2}(p-1)}$. It says

$$A^{p^{n-2}(p-1)} \not\equiv 1 \pmod{p^n}.$$

On the other hand,

$$A^{p^{n-2}(p-1)} \equiv 1 \pmod{p^{n-1}},$$

since $\varphi(p^{n-1}) = p^{n-2}(p-1)$. Thus $A^{p^{n-2}(p-1)}$ differs from 1 by a multiple of p^{n-1}) but not of p^n. Let

(##) $A^{p^{n-2}(p-1)} = 1 + tp^{n-1},$

where $p \nmid t$.

Since we are assuming the order of A modulo p^{n+1} is only $p^{n-1}(p-1)$, we will get our contradiction if we can show $A^{p^{n-1}(p-1)} \not\equiv 1 \pmod{p^{n+1}}$. The natural thing is to raise both sides of (##) to the power p. Doing this, and again applying Proposition (356), yields

$$A^{p^{n-1}(p-1)} = (1 + tp^{n-1})^p \equiv 1 + ptp^{n-1} \pmod{p^{2(n-1)}}.$$

If $2(n-1) \geqslant n+1$, this implies

$$A^{p^{n-1}(p-1)} \equiv 1 + p^n t \pmod{p^{n+1}}.$$

Thus $A^{p^{n-1}(p-1)} \not\equiv 1 \pmod{p^{n+1}}$, the desired contradiction.

We should verify that $2(n-1) \geqslant n+1$. This inequality is equivalent to $n \geqslant 3$. Thus our proof shows that all primitive roots modulo p^3 are primitive roots modulo p^4, that all primitive roots modulo p^4 are primitive roots modulo p^5, etc. But we have never proved that p^3 has a primitive root. We've taken care of p and p^2 but not p^3. There is a gap in our induction.

(560)   Let us spotlight the stubborn case of going from p^2 to p^3 by substituting $n = 2$ into the general discussion above. We assume A is a primitive root modulo p^2 and want to show it is also one modulo p^3. We know A has order $p(p-1)$ or $p^2(p-1)$ modulo p^3; since the latter is what we want, we assume the former and look for a contradiction. We also know A^{p-1} has the form $1 + tp$, where $p \nmid t$. Thus

$$A^{p(p-1)} = (1 + tp)^p,$$

and the argument would be complete if we could show the right side not congruent to 1 modulo p^3. But the Binomial Theorem, Exercise (358), says

$$(1 + tp)^p = 1 + ptp + \frac{p(p-1)}{2}(tp)^2 + (\text{terms with } p^3 \text{ as a factor}).$$

Thus

$$A^{p(p-1)} \equiv 1 + tp^2 \not\equiv 1 \pmod{p^3},$$

contradicting the assumption that the order of A is $p(p-1)$. We have proved

(561)   **Proposition.** If p is an odd prime and $\beta \geqslant 2$, then p^β has a primitive root. The primitive roots modulo p^β are exactly the primitive roots modulo p^2.

(562)   **Exercise.** Show by induction on n that

$$(1 + tm)^n \equiv 1 + ntm + \frac{n(n-1)}{2} t^2 m^2 \pmod{m^3}$$

for all positive integers n.

(563)   **Exercise.** Use (562) to show that

$$(1 + tm)^m \equiv 1 + tm^2 \pmod{m^3}$$

if m is odd.

56   MODULI HAVING PRIMITIVE ROOTS

(564)   The proof leading up to (561) was certainly a strange one. Most induction proofs involve a special argument for $n = 1$, often trivial, then a demonstration that the case n implies $n + 1$. In proving that p^n always has a primitive root the case $n = 1$ was not at all trivial. Then a special argument was needed to go from $n = 1$ to $n = 2$. Then *another* special argument was needed to go from $n = 2$ to $n = 3$.

The cynical reader may feel that the author is being excessively pedantic, and that any proposition that works for $n = 1$, for $n = 2$, and for $n + 1$, given its truth for $n \geqslant 3$, is bound to work for $n = 3$. Let him reflect that 2^1 has a primitive root, that 2^2 has a primitive root, and that the proof in (559) that a primitive root modulo p^n is also a primitive root modulo p^{n+1} for $n \geqslant 3$ is perfectly valid for $p = 2$. Only in going from p^2 to p^3 does the argument given break down for $p = 2$.

(565) **Exercise.** Where?

(566) **Exercise.** Find the smallest positive primitive root modulo m, if any, for
$m = 243, m = 343, m = 512$.

(567) Proposition (526) tells us that the integers $2p^\beta$, where p is an odd prime, also have primitive roots. Thus the doubtful cases are all cleared up, and we can combine all our results into

(568) **THEOREM.** The moduli m having primitive roots are exactly 2, 4, and those of the form p^β or $2p^\beta$, p any odd prime.

(569) **Exercise.** If n is a positive integer, let $P(n)$ be the number of integers m, $2 \leqslant m \leqslant n$, having primitive roots. Compute $P(10), P(20)$, and $P(50)$.

(570) **Exercise.** Show there exists an integer M such that $P(n)/n < 3/4$ for all $n \geqslant M$, where $P(n)$ is as in (569).

(571) **Exercise.** Guess whether the following is true or not: Given $\epsilon > 0$ there exists M such that $P(n)/n < \epsilon$ for all $n \geqslant M$, where $P(n)$ is as in (569).

57 THE PERIOD OF A DECIMAL EXPANSION

(572) If the decimal expansion of $1/m$ is computed, $m > 1$, the digits will be found to repeat after some point. Consider $1/7$, for example,

$$
\begin{array}{r}
.14285714\ldots \\
7 \overline{)\ 1.0000000} \\
7 \\
\hline
③0 \\
2\ 8 \\
\hline
②0 \\
1\ 4 \\
\hline
⑥0 \\
5\ 6 \\
\hline
④0 \\
3\ 5 \\
\hline
⑤0 \\
4\ 9 \\
\hline
①0 \\
7 \\
\hline
③\ .
\end{array}
$$

We know the decimal repeats because each succeeding digit of the quotient depends only on the previous remainder, and the remainders repeat. We have circled the remainders in the above calculation; they are 3, 2, 6, 4, 5, 1, 3,.... Since each remainder r satisfies $0 \leqslant r < 7$, there are at most 7 of them. Thus we could have predicted that the digits of the decimal for $1/7$ would repeat in groups of not more than 7.

(573) **Definition.** By $.a_1 a_2 \ldots a_j \overline{b_1 b_2 \ldots b_k}$ we mean the decimal $.a_1 a_2 \ldots a_j b_1 b_2 \ldots b_k b_1 b_2 \ldots b_k b_1 b_2 \ldots$. By the *period* of a number ξ we mean the smallest integer k such that $\xi = .a_1 a_2 \ldots a_j \overline{b_1 b_2 \ldots b_k}$.

(574) **Examples.** We have $1/6 = .1666\ldots = .1\overline{6}$, $1/7 = .\overline{142857}$, and $1/11 = .090909 \ldots = .\overline{09}$. The periods of $1/6$, $1/7$, and $1/11$ are 1, 6, and 2, respectively.

(575) **Exercise.** Express in the above notation $1/3$, $1/9$, and $1/13$. What are the periods?

(576) Since the remainders r in the long-division calculation of $1/m$ always satisfy $0 \leqslant r < m$, the period of $1/m$ can be at most m. If $(m, 10) = 1$, however, we can say even more. Imagine stopping the computation at some point. With $m = 7$, for example, consider the situation after two steps:

$$
\begin{array}{r}
.1\ 4 \\
7\overline{)1.0\ 0} \\
\underline{7} \\
3\ 0 \\
\underline{2\ 8} \\
2.
\end{array}
$$

By shifting the decimal points two places to the right we see this says

$$\frac{100}{7} = 14 + \frac{2}{7},$$

or $100 = 7 \cdot 14 + 2$.

In general, stopping the calculation of $1/m$ at the nth step gives us

$$\frac{10^n}{m} = q + \frac{r}{m},$$

where r is the remainder. This means

$$10^n = qm + r, \quad 0 \leqslant r < m.$$

We see $10^n \equiv r \pmod{m}$. Since $(m, 10) = 1$ implies $(m, 10^n) = 1$, we also have $(r, m) = 1$. Thus r can take on at most $\varphi(m)$ values, and the period of $1/m$ cannot exceed $\varphi(m)$.

(577) **Exercise.** Suppose m and k are integers, with $0 < k \leqslant m$ and $(10, m) = 1$. Show that the period of k/m does not exceed $\varphi(m)$.

(578) When does the decimal for $1/m$ first repeat? The first time we get the remainder 1, since this puts us back at the start. Thus the period is g if $n = g$ is the smallest positive solution to

$$10^n \equiv 1 \pmod{m}.$$

But g has another name; it is just the order of 10 modulo m.

(579) **Proposition.** Suppose m is a positive integer and $(m, 10) = 1$. Then the period of the decimal expansion of $1/m$ is exactly the order of 10 modulo m.

(580) In particular, the period must be a divisor of $\varphi(m)$. This knowledge may shorten the calculation of the period.

Suppose, for example, we didn't know the period of $1/7$. Since $\varphi(7) = 6$, 10 must have order 1, 2, 3, or 6 modulo 7. Now $10^2 = 100 \equiv 2$ and $10^3 \equiv 20 \equiv -1 \pmod 7$. Thus the order of 10 modulo 7 must be 6, and this is the period of $1/7$.

Taking $m = 17$ shows even better the power of this method. The period of $1/17$ must be 1, 2, 4, 8, or 16, since these are the divisors of $\varphi(17) = 16$. Now $10^2 = 100 \equiv -2 \pmod{17}$. Successively squaring both sides of this congruence yields $10^4 \equiv 4$ and $10^8 \equiv 16 \pmod{17}$. Thus the period must be 16.

(581) **Exercise.** Compute the decimal expansion of $1/17$.

(582) **Exercise.** Determine the period of $1/19$, both by long division and by the method of (580).

(583) **Exercise.** Determine the period of $1/21$.

(584) **Exercise.** Determine the period of $1/59$.

58 DECIMALS WITH MAXIMAL PERIOD

(585) Special attention has been paid in the past to the integers m such that $1/m$ has period $m - 1$. Since the period of $1/m$ cannot exceed $\varphi(m)$, this implies $m - 1 \leqslant \varphi(m)$. But we know $\varphi(m)$ is as big as $m - 1$ only for m prime.

Even for primes p, $1/p$ may not have period $p - 1$. This occurs only when 10 has order $p - 1$ modulo p; i.e., when 10 is a primitive root modulo p.

(586) **Proposition.** Suppose $(m, 10) = 1$. The decimal expansion of $1/m$ has period $m - 1$ if and only if m is prime and 10 is a primitive root modulo m.

(587) **Exercise.** Determine all integers m, $(m, 10) = 1$, such that $1/m$ has period $m - 1$ and $40 \leqslant m \leqslant 50$.

(588) Unfortunately the determination of the numbers m such that $1/m$ has has period $m - 1$ more or less dies with Proposition (584). No one knows even whether the set of primes having 10 as a primitive root is infinite or not.

(589) **Exercise.** Find all integers m, $(m,10) = 1$, such that $1/m$ has order 2.

(590) **Exercise.** Find all integers m, $(m,10) = 1$, such that $1/m$ has order 3.

(591) **Exercise.** Suppose $(10,m) = 1$. Show there exists an integer k, the decimal representation of which consists only of 9's, such that $m|k$.

(592) **Exercise.** Suppose $(10,m) = 1$. Show there exists an integer k, the decimal representation of which consists only of 1's, such that $m|k$.

(593) **Exercise.** Prove that if $(m,k) = 1$, $(m,10) = 1$, and $0 < k \leqslant m$, then the period of k/m is the same as the period of $1/m$.

59 COMPUTATION WITH INDICES

(594) Certain calculations in a reduced residue system modulo m may be simplified when m has a primitive root. The integer 2 is a primitive root modulo 11, for example. Consider the following table, where the powers of 2 have been replaced by the least positive integers congruent to them modulo 11.

n	0	1	2	3	4	5	6	7	8	9
2^n	1	2	4	8	5	10	9	7	3	6

The table says $2^4 \equiv 5$ (mod 11), $2^6 \equiv 9$ (mod 11), etc.

Suppose we wish to calculate the least positive integer congruent to $4 \cdot 7$ modulo 11. We note that

$$4 \cdot 7 \equiv 2^2 2^7 = 2^9 \equiv 6 \text{ (mod 11)}.$$

This looks longer than a direct calculation, but it is not really necessary to write down so much. To multiply two numbers we merely find them in the bottom row of the table, add the corresponding entries in the top row, and then find the entry below the sum back in the bottom row. If the sum exceeds 9, we subtract 10 from it first, since

$$2^{10} \equiv 2^0 \text{ (mod 11)}.$$

Thus to multiply 6 and 5, we find the entries 9 and 4 in the top row. Their sum is 13, which we reduce to 3. This corresponds to 8 in the bottom row. Thus $6 \cdot 5 \equiv 8$ (mod 11). [*Justification:* $6 \cdot 5 \equiv 2^9 2^4 = 2^{13} = 2^{10} 2^3 \equiv 2^3 \equiv 8$ (mod 11).]

We see multiplication is reduced to addition by the use of our table. The reader would probably not be surprised if we called the entries in its top row the "logarithms" of those below, but such is not traditional.

(595) **Definition.** Suppose A is a primitive root modulo m and $(k,m) = 1$. We call i the *index of k modulo m relative to A* in case $0 \leqslant i < \varphi(m)$, and

$$A^i \equiv k \pmod{m}.$$

(596) The restriction $0 \leqslant i < \varphi(m)$ in the definition above is merely to assure that each integer relatively prime to m has a unique index.

Suppose a and b have indices i and j respectively. Then

$$ab \equiv A^i A^j \equiv A^{i+j} \pmod{m}.$$

We cannot say that $i + j$ is the index of ab, since $i + j$ may be $\varphi(m)$ or greater. In this case

$$ab \equiv A^{\varphi(m)} A^{i+j-\varphi(m)} \equiv A^{i+j-\varphi(m)} \pmod{m},$$

and $i + j - \varphi(m)$ is the index of ab.

(597) **Proposition.** Suppose a and b are relatively prime to m, and that a, b, and ab have respectively the indices i_a, i_b, and i_{ab} modulo m relative to some primitive root. Then

$$i_{ab} \equiv i_a + i_b \pmod{\varphi(m)}.$$

(598) If $(a,m) = 1$, it is natural to denote by a^{-1} any solution to $ax \equiv 1 \pmod{m}$. If a and x have indices i_a and i_x, then $i_a + i_x \equiv 0 \pmod{\varphi(m)}$, since the index of 1 is 0. Thus

$$i_x \equiv -i_a \pmod{\varphi(m)}.$$

Consider, for example, the congruence $9x \equiv 1 \pmod{11}$. As in (594) our indices will be relative to the primitive root 2, and we can use the table there. Since the index of 9 is 6, the index of 9^{-1} is congruent to -6 modulo 10. It must be 4, which is the index of 5. Sure enough,

$$9 \cdot 5 \equiv 1 \pmod{11}.$$

We might write $9^{-1} \equiv 5 \pmod{11}$.

Congruences of the form $ax \equiv b \pmod{m}$ can also be solved this way when a and b are relatively prime to m. Consider

$$5x \equiv 6 \pmod{11}.$$

We write this as

$$x \equiv 6 \cdot 5^{-1} \pmod{11}.$$

Thus

$$\text{(index of } x) \equiv \text{(index of 6)} - \text{(index of 5)} = 9 - 4 = 5 \pmod{10}.$$

We see $x = 10$.

We leave the proof of the following proposition to the reader.

(599) **Proposition.** Suppose a and b have indices i_a and i_b modulo m. Then x is a solution to

$$ax \equiv b \ (\mathrm{mod} \ m)$$

if and only if the index of x is congruent to $i_b - i_a$ modulo $\varphi(m)$.

(600) **Exercise.** Write out a table of indices such as the one in (594) for the modulus 26, using the primitive root 15. What is the index of 19? Of 9?

(601) **Exercise.** Find x satisfying each of the following congruences, $1 \leqslant x \leqslant 26$.
  (a) $17 \cdot 23 \equiv x \ (\mathrm{mod} \ 26)$
  (b) $5 \cdot 19 \equiv x \ (\mathrm{mod} \ 26)$
  (c) $31 \cdot 41 \equiv x \ (\mathrm{mod} \ 26)$
  (d) $7x \equiv 1 \ (\mathrm{mod} \ 26)$
  (e) $9x \equiv 21 \ (\mathrm{mod} \ 26)$.

chapter 7

Quadratic Reciprocity

The problem of distinguishing quadratic residues from nonresidues will be answered with a proof of the celebrated Law of Quadratic Reciprocity of K. F. Gauss.

60 THE INDEX OF A QUADRATIC RESIDUE

(602) Suppose a has index i modulo m relative to the primitive root A. Then the index of a^2 is congruent to $i + i = 2i$ modulo $\varphi(m)$. If $m > 2$ then $\varphi(m)$ is even, so

$$(\text{index of } a^2) \equiv 2i \ (\text{mod } 2).$$

Thus the index of a^2 is even.

Conversely, if the index of b is $2j$, then

$$b \equiv (A^j)^2 \ (\text{mod } m).$$

(603) **Proposition.** If $m > 2$ has a primitive root, then any integer relatively prime to m is a quadratic residue modulo m if and only if its index is even.

(604) **Example.** From the table in (594) we see that 1, 4, 5, 9, and 3 are quadratic residues modulo 11, while 2, 8, 10, 7, and 6 are nonresidues.

(605) **Exercise.** What are the quadratic residues a modulo 26, $1 \leqslant a \leqslant 26$?

(606) **Exercise.** Find all integers a, $1 \leqslant a \leqslant 26$, $(a, 26) = 1$, such that the congruence $x^3 \equiv a \ (\text{mod } 26)$ is solvable. What are their indices?

(607) **Exercise.** What are the indices of all the integers a, $(a, 43) = 1$, such that $x^3 \equiv a \ (\text{mod } 43)$ is solvable?

(608) **Exercise.** What are the indices of all the integers a, $(a, 11) = 1$, such that $x^3 \equiv a \ (\text{mod } 11)$ is solvable?

(609) **Exercise.** Prove that if m has a primitive root, $k \mid \varphi(m)$, and $(a, m) = 1$, then $x^k \equiv a \ (\text{mod } m)$ has a solution if and only if k divides the index of a.

(610)   Quadratic residues have even indices; nonresidues have odd indices; to multiply numbers modulo m we add their indices—but the reader has probably guessed the next result already.

(611) **Proposition.** The product of two quadratic residues or of two non-residues modulo m is a quadratic residue. The product of a quadratic residue and a nonresidue is a nonresidue.

Proof. Suppose a and b are quadratic residues modulo m. Let the index of a modulo m relative to some primitive—whoops! We forgot to assume m had a primitive root, and indices were only defined for such m.

Of course if the author had really forgotten he would have gone back and inserted the necessary hypothesis, with no one the wiser. Why didn't he? Because the hypothesis is *not* necessary; the proposition is valid for any modulus m.

(612) **Exercise*.** Prove Proposition (611) in two ways: (1) directly from the definition of quadratic residue and nonresidue, (2) using (462).

(613) **Exercise.** Suppose $k > 1$ and a and b are relatively prime to m. Consider

$$x^k \equiv a \pmod{m}$$

$$x^k \equiv b \pmod{m}$$

$$x^k \equiv ab \pmod{m}.$$

Show that if any two of these congruences have solutions, then so does the third.

61 THE LEGENDRE SYMBOL

(614)   Any prime p has a primitive root, so Proposition (603) may be used to determine its quadratic residues; but we cannot pretend that this is an efficient method, especially when the only question is whether a single integer is a quadratic residue or not. To tell whether 13 is a quadratic residue modulo the prime 31 this way we would first have to find a primitive root modulo 31, which might entail several false starts. Better to apply Theorem (462), which says that if p is an odd prime and $p \nmid a$, then a is a quadratic residue modulo p or not according as $a^{(p-1)/2}$ is congruent to 1 or -1 modulo p.

(615) **Exercise.** Is 13 a quadratic residue modulo 31? Use any method.

(616) **Definition.** Suppose p is an odd prime and $p \nmid a$. We define (a/p), called the *Legendre symbol*, to be 1 if a is a quadratic residue modulo p and -1 if a is a nonresidue.

(617) **Examples.** Since 1 is a quadratic residue modulo 3 and 2 is not we have $(1/3) = 1$ and $(2/3) = -1$. Also $(16/3) = 1$, $(-1/3) = -1$, $(1/5) = (4/5) = 1$, $(2/5) = (3/5) = -1$. At present $(6/3)$, $(1/2)$, $(2/15)$, and $(5/9)$ are all undefined.

(618) **Exercise.** Evaluate each of the following (some may be undefined): $(8/7), (9/7), (9/8), (100/3), (-14/7), (5/21), (6/11), (-1/2), (-1/31)$.

(619) Previous results show that the Legendre symbol has many pleasant characteristics (see (462), (469), and (611), for example). These are collected in the following proposition, which the reader will have the honor of proving.

(620) **Proposition.** Suppose p is an odd prime not dividing a or b. Then

    (a) $(a^2/p) = 1$
    (b) $(ab/p) = (a/p)(b/p)$
    (c) $(a/p) \equiv a^{(p-1)/2} \pmod{p}$
    (d) If $a \equiv b \pmod{p}$, then $(a/p) = (b/p)$
    (e) $(-1/p) = 1$ or -1 according as $p \equiv 1$ or $3 \pmod{4}$.

(621) Part (b) above is especially interesting because it says that, regarded as a function of two variables, the symbol (a/p) is multiplicative in the first. Better than just multiplicative, in fact, since $(ab/p) = (a/p)(b/p)$ whether $(a, b) = 1$ or not.

    Suppose we want to evaluate $(20/31)$. We know $(20/31) = (2^2/31)(5/31)$. Now $(2^2/31) = 1$ by part (a) above, so the problem is reduced to determining $(5/31)$. Unfortunately computing $5^{(31-1)/2} \pmod{31}$ is not particularly easier than $20^{(31-1)/2} \pmod{31}$. If we had a practical method of computing (a/p) for prime a, we would be in good shape. We do not.

62 GAUSS'S LEMMA

(622) Theorem (462) is the only result of any depth we have concerning quadratic residues. The trouble with it is that our method of computing $a^{(p-1)/2} \pmod{p}$ is more or less brute force. We need a better way.

    We evaluated a similar expression once, in proving Fermat's Theorem. There we found that

$$a^{p-1} \equiv 1 \pmod{p}.$$

Our method of proof turned out to be adaptable to other theorems, among them (462) itself.

    Let us review the idea once more. We let R be a reduced residue system modulo p. Then as r runs through R, ra also runs through a reduced residue system modulo p. Thus

$$a^{p-1} \prod_{r \in R} r = \prod_{r \in R} ra \equiv \prod_{r \in R} r \pmod{p}.$$

Canceling $\prod_{r \in R} r$ from both sides gives Fermat's Theorem.

    What can we do like this with $a^{(p-1)/2}$? We must somehow use a product over $(p-1)/2$ instead of p factors. This presents no problem; instead of a reduced residue system modulo p we could use the set $S = \left\{1, 2, \cdots, \dfrac{p-1}{2}\right\}$, for instance. The set S has exactly $(p-1)/2$ elements. The trouble comes with $\Pi_{s \in S}\ sa$. The numbers sa are all distinct, but they are not necessarily all congruent to elements of S, since S is not a reduced residue system.

There *is* a reduced residue system the elements of which are closely related to those of S, namely the set S^* pictured below.

S^* (0 not included)

The set S^* is formed by omitting 0 from the p consecutive integers from $-\dfrac{p-1}{2}$ to $\dfrac{p-1}{2}$, and is clearly a reduced residue system modulo p. We can also think of S^* as comprising the elements of S along with their negatives. It is natural to call the latter integers $-S$.

Given s in S there exists a unique element s^* in S^* such that

$$sa \equiv s^* \pmod{p}.$$

Furthermore, the absolute value of s^* is back in S, which is the idea.

Let us consider $p = 7$ and $a = 2$. Here $S = \{1, 2, 3\}$ and $S^* = \{-3, -2, -1, 1, 2, 3\}$. Note that $1 \cdot 2 \equiv 2$, $2 \cdot 2 = 4 \equiv -3$, and $3 \cdot 2 = 6 \equiv -1 \pmod{7}$. Thus we have

s	1	2	3
s^*	2	-3	-1
$\lvert s^* \rvert$	2	3	1

Notice that in this case as s runs through S so does $\lvert s^* \rvert$. We will use this fact to evaluate $(2/7)$, and what we do will illustrate the general theorem to come.

From (620c) we know that $(2/7) \equiv 2^3 \pmod{7}$. Now $\Pi_{s \in S}\ s \cdot 2 = 2^3\ \Pi_{s \in S}\ s$, since S has just 3 elements. On the other hand,

$$\prod_{s \in S} s \cdot 2 \equiv \prod_{s \in S} s^* = (2)(-3)(-1) = 1 \cdot 2 \cdot 3 = \prod_{s \in S} s \pmod{7}.$$

Thus

$$2^3 \prod_{s \in S} s \equiv \prod_{s \in S} s \pmod{7}.$$

We cancel $\prod_{s \in S} s$ to get $2^3 \equiv 1 \pmod 7$. Thus $(2/7) = 1$.

(623) **Exercise.** Make a table of s, s^*, and $|s^*|$ as above for $p = 11$ and $a = 3$. Use it to evaluate $(3/11)$.

(624) **Exercise.** Make a table of s, s^* and $|s^*|$ for $p = 13$ and $a = 5$. Use it to evaluate $(5/13)$.

(625) In our example above we wound up canceling $\prod_{s \in S} s$ from both sides of a congruence. We *had* $\prod_{s \in S} s$ on the right-hand side because we found that as s ran through S, so did $|s^*|$.

(626) **Proposition.** Suppose p is an odd prime and $p \nmid a$. Given s, let s^* be the unique integer satisfying $-\dfrac{p-1}{2} \leqslant s^* \leqslant \dfrac{p-1}{2}$ such that $sa \equiv s^* \pmod p$. Then as s runs through the numbers $1, 2, \ldots, (p-1)/2$, so does $|s^*|$.

Proof. Since the numbers $|s^*|$ all fall back into S, it suffices to show they are distinct. We will assume that s_1 and s_2 are in S and $|s_1^*| = |s_2^*|$, and show $s_1 = s_2$.

If s_1^* and s_2^* are both positive or both negative, then $s_1^* = s_2^*$. Thus

$$s_1 a \equiv s_1^* = s_2^* \equiv s_2 a \pmod p.$$

Since $p \nmid a$, we conclude $s_1 \equiv s_2 \pmod p$. Thus $s_1 = s_2$.

Otherwise we have $s_1^* = -s_2^*$. Then $s_1 a \equiv -s_2 a \pmod p$. This says $p \,|\, (s_1 + s_2)a$. We see $p \,|\, s_1 + s_2$. But by the definition of S,

$$0 < s_1 + s_2 \leqslant \frac{p-1}{2} + \frac{p-1}{2} = p - 1.$$

This is impossible.

(627) Now let us see what happens when we try to do in general what we did with $p = 7$ in (622). We have $(a/p) \equiv a^{(p-1)/2} \pmod p$ by Theorem (620). We note that

$$\prod_{s \in S} sa = a^{(p-1)/2} \prod_{s \in S} s,$$

since S has exactly $(p-1)/2$ elements. Thus

$$a^{(p-1)/2} \prod_{s \in S} s = \prod_{s \in S} sa \equiv \prod_{s \in S} s^* \pmod p,$$

by the definition of s^*. We must somehow replace each s^* with $|s^*|$ in order to use the last proposition.

We know $s^* = \pm|s^*|$, where we have the plus sign if s^* is in S and the minus sign if s^* is in $-S$. Suppose s^* is negative for exactly n values as s runs through S. Then

$$a^{(p-1)/2} \prod_{s \in S} s \equiv \prod_{s \in S} s^* = (-1)^n \prod_{s \in S} |s^*| = (-1)^n \prod_{s \in S} s \pmod{p},$$

by Proposition (626). We cancel $\prod_{s \in S} s$ to get

$$a^{(p-1)/2} \equiv (-1)^n \pmod{p}.$$

(628) **THEOREM.** (Gauss's Lemma)   Suppose p is an odd prime and $p \nmid a$. Suppose exactly n of the integers $1 \cdot a, 2 \cdot a, \cdots, \dfrac{p-1}{2} \cdot a$ are congruent to any of the numbers $-1, -2, \cdots, -\dfrac{p-1}{2}$ modulo p. Then $(a/p) = (-1)^n$.

(629) **Example.** We will evaluate $(^6/_{31})$. Since $(31 - 1)/2 = 15$, we look at the integers $s \cdot 6$, $1 \leqslant s \leqslant 15$.

s	1	2	3	4	5	6	7	8	9	10	11	12	13	14	15
s^*	6	12	-13	-7	-1	5	11	-14	-8	-2	4	10	-15	-9	-3

We count the minus signs and find $n = 9$. Thus $(^6/_{31}) = -1$.

(630) **Exercise.** Make a table of s and s^* for $p = 23$ and $a = 7$. What is n? What is $(^7/_{23})$?

(631) **Exercise.** Make a table of s and s^* for $p = 23$ and $a = 2$. What is n? What is $(^2/_{23})$?

(632) **Exercise.** Let $p = 101$ and $a = 2$. What is n? What is $(^2/_{101})$?

(633) *Note.* The reader is strongly advised to work the last two exercises before proceeding, since they illustrate our next result.

63 EVALUATING $(2/p)$

(634) Although using Gauss's Lemma to determine quadratic residues is no easier than any other method, this result will turn out to have theoretical importance.

Let us see what it tells us about the simplest case, $a = 2$. In order to determine whether 2 is a quadratic residue modulo p we are supposed to examine the integers $1 \cdot 2,\ 2 \cdot 2, \cdots, \dfrac{p-1}{2} \cdot 2$, and count the number congruent modulo p to elements of $-S$. The integers in question are just the even numbers from 2 to $p-1$; and those congruent to elements of $-S$ are easily seen to be just those exceeding $p/2$.

(635) **Exercise***. Suppose $k \equiv s^*$ (mod p), where s^* is in S^* and $1 \leqslant k \leqslant p-1$. Show s^* is in $-S$ if and only if $k > p/2$.

(636) **Exercise***. Suppose M and N are integers and $M \leqslant N$. Show that the number of integers s satisfying $M \leqslant s \leqslant N$ is exactly $N - M + 1$.

(637) We return to the problem of evaluating $(2/p)$. We have seen that all we have to do is count the even integers between $p/2$ and p. This amounts to counting the integers s such that

$$p/2 < 2s < p,$$

or

$$p/4 < s < p/2.$$

We can use Exercise (636) to do this if we determine the least and greatest of the integers we wish to count.

The number $p/4$ is not an integer, and how far shy it is of the next integer depends on p. The next integer to $3/4$ is $3/4 + 1/4 = 1$, while the next integer to $5/4$ is $5/4 + 3/4 = 2$. We must consider cases.

Case I. $p \equiv 1$ (mod 4).

Let $p = 4t + 1$. Then $p/4 = \dfrac{4t+1}{4} = t + 1/4$. The next integer is clearly

$$p/4 + 3/4 = \frac{p+3}{4}.$$

Case II. $p \equiv 3$ (mod 4).

Let $p = 4u + 3$. Then $p/4 = \dfrac{4u+3}{4} = u + 3/4$. The next integer is $p/4 +$

$$1/4 = \frac{p+1}{4}.$$

Recall we are counting the integers s such that

$$p/4 < s < p/2.$$

Since p is odd the integer previous to $p/2$ is $p/2 - \frac{1}{2} = \dfrac{p-1}{2}$. Thus in Case I the inequality is equivalent to

$$\frac{p+3}{4} \leqslant s \leqslant \frac{p-1}{2},$$

and (636) tells us

$$n = \frac{p-1}{2} - \frac{p+3}{4} + 1 = \frac{p-1}{4}.$$

Whether $(p-1)/4$ is even or odd depends on whether $8\,|\,p-1$ or not. Since $p \equiv 1 \pmod 4$ we have $p \equiv 1$ or $5 \pmod 8$. We see n is even if $p \equiv 1 \pmod 8$ and odd if $p \equiv 5 \pmod 8$.

In Case II we must count the integers s such that

$$\frac{p+1}{4} \leqslant s \leqslant \frac{p-1}{2}.$$

There are $\dfrac{p-1}{2} - \dfrac{p+1}{4} + 1 = \dfrac{p+1}{4}$ of them. This is even if and only if $p \equiv -1 \pmod 8$. Note that $p \equiv 3 \pmod 4$ implies $p \equiv 3$ or $p \equiv 7 \equiv -1 \pmod 8$.

(638) **Proposition.** Suppose p is an odd prime. Then $(2/p) = 1$ if and only if $p \equiv \pm 1 \pmod 8$.

(639) **Exercise.** Evaluate $(2/5)$, $(2/41)$, $(2/43)$, and $(2/101)$.

(640) **Exercise.** Evaluate $(8/37)$, $(-2/53)$, $(18/47)$, and $(39/37)$.

(641) **Exercise.** What is the greatest integer s such that $s < (p+3)/4$ for $p \equiv 1, 3, 5,$ and $7 \pmod 8$?

(642) **Exercise.** Suppose $p \equiv 3 \pmod 8$. How many integers s satisfy $(p-1)/4 \leqslant s \leqslant (p+2)/2$?

(643) **Exercise.** Give a proof of (620)(e) based on Gauss's Lemma.

(644) **Exercise.** Is 2 a quadratic residue modulo 77? Modulo 1001? Modulo 323? Modulo 833? [*Hint*: See (426).]

(645) **Exercise.** Show that if p is an odd prime, then $(2/p) = (-1)^{(p^2-1)/8}$.

64 EVALUATING $(3/p)$

(646) Our success in finding a general rule for evaluating $(2/p)$ emboldens us to try the case $a = 3$. Assume $p > 3$. Again we must determine how many of the integers

$$1 \cdot 3, \, 2 \cdot 3, \, \ldots, \, \frac{p-1}{2} \cdot 3$$

are congruent to elements of $-S$ modulo p. These are just the positive multiples of 3 up to $3(p-1)/2$, and the number n we desire merely counts the number of integers s such that

$$p/2 < 3s < p,$$

or

$$p/6 < s < p/3.$$

The extreme values of s depend on what p is modulo 6. Since $p > 3$, $p \equiv 1$ or 5 (mod 6).

Case I. $p \equiv 1$ (mod 6).

Let $p = 6t + 1$. Then $\dfrac{p}{6} = \dfrac{6t+1}{6} = t + \dfrac{1}{6}$. The next integer to the right is

$\dfrac{p}{6} + \dfrac{5}{6} = \dfrac{p+5}{6}$. Also $\dfrac{p}{3} = \dfrac{6t+1}{3} = 2t + \dfrac{1}{3}$. The nearest integer to the left of $\dfrac{p}{3}$ is

$\dfrac{p}{3} - \dfrac{1}{3} = \dfrac{p-1}{3}$. We see $n = \dfrac{p-1}{3} - \dfrac{p+5}{6} + 1 = \dfrac{p-1}{6}$. Now $p \equiv 1$ or 7 (mod 12)

and n is even only in the former case.

We leave Case II to the reader.

(647) **Exercise*.** Continue the above proof for the case $p \equiv 5$ (mod 6). Show n is even if $p \equiv -1$ (mod 12) and odd if $p \equiv 5$ (mod 12).

(648) **Exercise*.** Show that if $p > 3$ is prime then $(3/p) = 1$ if and only if $p \equiv \pm 1$ (mod 12).

(649) **Exercise.** Evaluate $(3/41)$, $(3/47)$, $(6/31)$, $(54/101)$, and $(-3/43)$.

(650) **Exercise.** Show that if $p > 3$ is prime, then $(3/p)(p/3) = (-1)^{(p-1)/2}$.

(651) **Exercise*.** Show that if p is an odd prime other than 5, then $(5/p) = (-1)^n$, where n is the total number of integers s satisfying

$$p/10 < s < p/5 \quad \text{or} \quad 3p/10 < s < 2p/5.$$

65 CONSIDERING (a/p) AS A FUNCTION OF p

(652) Our applications of Gauss's Lemma, while successful, are getting messier and messier. According to the last exercise, for example, we must count the integers in *two* intervals in order to evaluate $(5/p)$. For bigger values of a we would expect even more intervals, since the range of the numbers sa, $1 \leqslant s \leqslant \dfrac{p-1}{2}$, increases with a.

One pattern has emerged. The value of $(2/p)$ depends on what p is modulo 8. Likewise $(3/p)$ depends on what p is modulo 12. A reasonable guess is that the value of (a/p) depends on what p is modulo $4a$, and our proofs for $a = 2$ and 3 give an inkling of *why* this should be so.

To get started with a proof let us consider $a = 5$, seeing if we can some-how show that $(5/p)$ depends only on what p is modulo 20 without going so far as to evaluate the symbol.

By (651) it suffices to count the integers satisfying

$$p/10 < s < p/5 \quad \text{or} \quad 3p/10 < s < 2p/5$$

to determine n. We want to show that if p changes by any multiple of 20 then the "parity" (i.e., evenness or oddness) of n doesn't change.

Let us substitute $p + 20$ for p in the first inequality. We get

$$\frac{p + 20}{10} < s < \frac{p + 20}{5}$$

or

$$\frac{p}{10} + 2 < s < \frac{p}{5} + 4.$$

We see that the left-hand endpoint of the interval defined by the inequality shifts 2 units to the right and the right-hand endpoint shifts 4 to the right. Since each endpoint moves an even number of units, *an even number of integers are gained or lost by the change.* (Actually 2 are lost and 4 gained, a net increase of 2.)

Likewise replacing p by $p + 20$ in the second inequality gives

$$\frac{3p}{10} + 6 < s < \frac{2p}{5} + 8,$$

and again each endpoint moves an even number of units.

We see that the parity of n is unchanged by replacing p by $p + 20$. Since this operation may be repeated indefinitely, we have proved that *if p and q are odd primes other than 5 and $p \equiv q$ (mod 20), then $(5/p) = (5/q)$.*

(653) **Exercise.** Evaluate $(5/43)$, $(10/23)$, and $(-5/47)$.

(654) The last proof involved counting the integers in certain intervals. We will use the standard notation (α, β) to denote the set of real numbers x satisfying $\alpha < x < \beta$. The following proposition legitimizes certain useful interval manipulations, most of them obvious from a picture. The reader should prove it.

(655) **Proposition.** Let α and β be real numbers, $\alpha < \beta$, and let k be an integer. Let n be the number of integers in (α, β).
   (a) The number of integers in $(\alpha + k, \beta)$ is $n - k$ if $\alpha + k < \beta$.
   (b) The number of integers in $(\alpha, \beta + k)$ is $n + k$ if $\alpha < \beta + k$.
   (c) The number of integers in $(-\beta, -\alpha)$ is n.
   (d) If $k > 0$ the number of integers j such that jk is in (α, β) is the same as the number of integers in $(\alpha/k, \beta/k)$.

(656) **Proposition.** If p and q are odd primes not dividing $a > 0$ and if $p \equiv q \pmod{4a}$, then $(a/p) = (a/q)$.

Proof. To evaluate (a/p) we must determine whether the number of integers s in $S = \{1, 2, \ldots, (p-1)/2\}$ such that sa is congruent to an element of $-S \pmod p$ is even or odd.

Note that the numbers $\frac{p}{2}, p, \frac{3}{2}p, 2p, \ldots$ divide the positive real axis into an infinite number of intervals:

The integers sa, $s \in S$, are just all multiples of a in the intervals up through the one with right-hand endpoint $\frac{a}{2}p$, since

$$\frac{p-1}{2} \cdot a < \frac{a}{2} \cdot p < \frac{p+1}{2} \cdot a.$$

We only want to count the multiples of a in about half of these intervals, namely, the ones of the form

$$((k - \tfrac{1}{2})p, kp), k = 1, 2, \ldots, K,$$

since these contain exactly the integers congruent to elements of $-S$. The number of such intervals, which we have called K, depends only on a; its exact value doesn't concern us in this proof.

We see n counts the total number of integers s satisfying

$$sa \in ((k - \tfrac{1}{2})p, kp)$$

for some k. This is equivalent to

$$s \in ((k - \tfrac{1}{2})p/a, kp/a),$$

by (655) (d).

Our proof will consist in showing that replacing p by q in the above interval does not change the parity of the number of integers it contains. We know $q \equiv p \pmod{4a}$; let $q = p + 4at$. Then replacing p by q changes the left endpoint of the interval by

$$\frac{(k - \tfrac{1}{2})q}{a} - \frac{(k - \tfrac{1}{2})p}{a} = \frac{(k - \tfrac{1}{2})(p + 4at - p)}{a} = (k - \tfrac{1}{2})4t = 2(2k - 1)t.$$

The right-hand endpoint changes by

$$\frac{kq}{a} - \frac{kp}{a} = \frac{k(p + 4at - p)}{a} = 4tk.$$

Since both of these numbers are even, the proof is completed.

(657) **Exercise**. Show that if $4a|p - q$, where p and q are distinct odd primes, then $(a/p) = (a/q)$ even if a is negative. [*Hint*: $(|a|/p) = (|a|/q)$.]

(658) **Exercise**. Evaluate $(7/31)$, $(11/47)$, and $(13/59)$.

66 QUADRATIC RECIPROCITY

(659) If p and q are odd primes, $p > q$, and if $p \equiv q$ (mod 4), then $p = q + 4a$ for *some* a, and so $(a/p) = (a/q)$. The integer $4a$ has the interesting property of being congruent to p (mod q) and congruent to $-q$ (mod p). Thus $(4a/p) = (-q/p)$ and $(4a/q) = (p/q)$.

But $(4a/q) = (4/q)(a/q) = (a/q)$ by parts (b) and (a) of Theorem (620). Likewise $(4a/p) = (a/p)$. We see

$$(p/q) = (4a/q) = (a/q) = (a/p) = (4a/p) = (-q/p).$$

We can do more with $(-q/p)$; it is $(-1/p)(q/p)$. We know $(-1/p)$ is 1 or -1 according as p [and therefore q; remember $p \equiv q$ (mod 4)] is congruent to 1 or 3 (mod 4).

(660) **Proposition**. Suppose p and q are distinct odd primes and $p \equiv q$ (mod 4). Then $(p/q) = (q/p)$ or $-(q/p)$ according as p and q are congruent to 1 or 3 modulo 4.

(661) The great value of the last result in calculation soon becomes apparent. Consider $(7/31)$, for example. Since $7 \equiv 31 \equiv 3$ (mod 4), it says $(7/31) = -(31/7) = -(3/7)$, where we have used (620) (d). In the same way $(3/7) = -(7/3) = -(1/3) = -1$. Thus $(7/31) = 1$.

(662) **Exercise**. Determine $(7/47)$, $(15/101)$, $(21/43)$, and $(11/41)$.

(663) If we had a result similar to Proposition (660) for primes *not* congruent modulo 4 we would really be in great shape. Let us see what we can come up with by working along the same lines as in (659).

Suppose one of p and q is congruent to 1 (mod 4) and the other to 3. Then $p + q = 4a$ for some positive integer a. We note $(a/p) = (4a/p) = (p + q/p) = (q/p)$ and $(a/q) = (4a/q) = (p + q/q) = (p/q)$.

We need something relating (a/p) and (a/q). Notice that $p \equiv -q$ (mod 4a). We know the value of (a/p) depends only on what p is modulo $4a$. The meager evidence provided by our results for $a = 2$ and $a = 3$ suggests, in fact, that if P is an odd prime then (a/P) is the same whether $P \equiv p$ or $-p$ (mod 4a).

Thus if $a = 2$, $4a = 8$; and we saw $(2/P) = 1$ if $P \equiv \pm 1$ (mod 8) and $(2/P) = -1$ if $P \equiv \pm 3$ (mod 8). For $a = 3$ the congruence classes modulo 12 show a similar symmetry. Let us attempt to prove

(664) **Proposition**. If p and q are odd primes and $p + q = 4a$, then $(a/p) = (a/q)$.

Proof. We will copy the proof given for Proposition (656). It suffices to show that replacing p by q in the interval

$$((k - \tfrac{1}{2})p/a, kp/a)$$

changes the number of integers in that interval by an even amount, if at all.

We know $q = 4a - p$. Thus such a replacement causes the right-hand endpoint to shift by the amount

$$\frac{(k - \tfrac{1}{2})q}{a} - \frac{(k - \tfrac{1}{2})p}{a} = \frac{(k - \tfrac{1}{2})(4a - p - p)}{a}$$

$$= \frac{(2k - 1)(2a - p)}{a} = 2(2k - 1) - \frac{p(2k - 1)}{a}$$

This may not even be an integer, much less even, since there is no reason to believe $a \mid 2k - 1$. We will have to be trickier than in the proof of (656).

Consider the new interval, after the replacement of p by q. It is

$$((k - \tfrac{1}{2})(4a - p)/a, k(4a - p)/a) = \left(4k - 2 - \frac{(k - \tfrac{1}{2})p}{a}, 4k - \frac{kp}{a}\right).$$

The numbers involved here are similar to those of our original interval except for certain minus signs. Let us use part (c) of Proposition (655) to replace this by the interval

$$\left(\frac{kp}{a} - 4k, \frac{(k - \tfrac{1}{2})p}{a} - 4k + 2\right),$$

which still contains the same number of integers.

Neither will the number of integers change if we add $4k - 2$ to each endpoint, producing

$$\left(\frac{kp}{a} - 2, \frac{(k - \tfrac{1}{2})p}{a}\right).$$

The original interval was $\left(\frac{(k - \tfrac{1}{2})p}{a}, \frac{kp}{a}\right)$. *Together* these two intervals comprise

$$\left(\frac{kp}{a} - 2, \frac{kp}{a}\right).$$

$$\frac{kp}{a} - 2 \qquad \frac{(k - \tfrac{1}{2})p}{a} \qquad \frac{kp}{a}$$

We must worry about whether we have picked up an extra integer by combining these intervals in this way; maybe $(k - \tfrac{1}{2})p/a$ was an integer. An examination of the origin of these intervals back in the proof of (656) shows that in all cases $k < a$ (in fact $k \leqslant a/2$). Since $p \neq q$, $p + q = 4a$ implies $p \nmid a$.

We see that not only is $(k - \frac{1}{2})p/a$ not an integer; neither is kp/a. Thus the interval

$$\left(\frac{kp}{a} - 2, \frac{kp}{a}\right)$$

contains exactly 2 integers. These may be divided between its two subintervals in only three ways: 0-2, 1-1, and 2-0. *In each case the parity of the number of intervals in one subinterval is the same as that of the number in the other.* This completes the proof.

(665) **THEOREM.** The Law of Quadratic Reciprocity. Suppose p and q are distinct odd primes. Then $(p/q) = (q/p)$ unless p and q are both congruent to 3 modulo 4, in which case $(p/q) = -(q/p)$.

Proof. If $p \not\equiv q \pmod 4$, then there exists an integer a such that $p + q = 4a$. Proposition (664) then says $(a/p) = (a/q)$. But $(a/p) = (q/p)$ and $(a/q) = (p/q)$ by the argument given in (663).

The case $p \equiv q \pmod 4$ is taken care of by Proposition (660).

(666) **Examples.** Let us compute $(^{13}/_{43})$. Since $13 \equiv 1 \pmod 4$ we have $(^{13}/_{43}) = (^{43}/_{13}) = (3 \cdot 13 + 4/13) = (^{4}/_{13}) = 1$, where we have used parts (d) and (a) of Proposition (620).

Let us compute $(^{19}/_{59})$. We have $(^{19}/_{59}) = -(^{59}/_{19}) = -(^{2}/_{19}) = -(-1) = 1$, where we used Proposition (638) to evaluate $(^{2}/_{19})$.

Likewise $(^{37}/_{67}) = (^{67}/_{37}) = (^{30}/_{37}) = (^{2}/_{37})(^{3}/_{37})(^{5}/_{37}) = (-1)(^{37}/_{3})(^{37}/_{5}) = -(^{1}/_{3})(^{2}/_{5}) = -(1)(-1) = 1$.

(667) **Exercise.** Evaluate $(^{17}/_{37})$, $(^{23}/_{43})$, $(^{44}/_{71})$, and $(^{71}/_{101})$.

(668) **Exercise.** Determine whether each of the first numbers of the following pairs is a quadratic residue modulo the second: 15, 41; 1000, 157; 234, 157; 158, 159.

(669) **Exercise.** Prove that the number K defined in the proof of Proposition (656) is $a/2$ if a is even and $(a - 1)/2$ if a is odd.

(670) **Exercise.** Show that if p and q are distinct odd primes, then $(p/q) \cdot (q/p) = (-1)^{(p-1)(q-1)/4}$.

67 THE JACOBI SYMBOL

(671) Calculation with the Legendre symbol (a/p) is hampered by the fact that a must be prime to use the reciprocity law. To determine $(105/211)$, for

example, we must invert $(3/211)$, $(5/211)$, and $(7/211)$ separately. We know
that

$$(105/211) = \pm(211/3)(211/5)(211/7);$$

let us see if we can discover some short way to tell whether the plus or minus
sign is correct.

Suppose $p_1, p_2, \ldots, p_t$, and q are odd primes, and let $P = p_1 p_2 \ldots p_t$.
We are looking at

$$(P/q) = (p_1 p_2 \cdots p_t/q) = (p_1/q)(p_2/q) \cdots (p_t/q).$$

If $q \equiv 1 \pmod 4$ everything is easy, for then $(p_i/q) = (q/p_i)$ for all i. Let us
suppose $q \equiv 3 \pmod 4$. Then we see

$$(p_1 p_2 \cdots p_t/q) = (-1)^n (q/p_1)(q/p_2) \cdots (q/p_t),$$

where n is just the number of the p's congruent to $-1 \pmod 4$. But since each p
is congruent to $\pm 1 \pmod 4$, $p_1 p_2 \cdots p_t \equiv (-1)^n \pmod 4$, where n is exactly the
same number mentioned in the last sentence.

We see that $(P/q) = (q/p_1)(q/p_2) \cdots (q/p_t)$ unless q and P are both congruent to 3 (mod 4), in which case $(P/q) = -(q/p_1)(q/p_2) \cdots (q/p_t)$.

This sounds exactly like the quadratic reciprocity law, except that we
must write $(q/p_1)(q/p_2) \cdots (q/p_t)$ instead of (q/P), since the latter is undefined.
An enlargement of our definition is indicated.

(672) **Definition.** Suppose $P = p_1 p_2 \cdots p_t$, where the p's are odd primes.
Suppose $(a,P) = 1$. We define (a/P), called the *Jacobi symbol*, to be

$$(a/p_1)(a/p_2) \cdots (a/p_t).$$

(673) Notice that if P is itself prime then (a/P) has its previous meaning. Thus
using the same notation as for the Legendre symbol is justified. Of course we
cannot just assume all the results we proved for the Legendre symbol also hold
for the Jacobi symbol. They must all be checked out again.

(674) **Proposition.** Suppose P and Q are relatively prime positive odd numbers.
Then $(P/Q) = (Q/P)$ unless $P \equiv Q \equiv 3 \pmod 4$, in which case $(P/Q) = -(Q/P)$.

Proof. Let $P = p_1 p_2 \cdots p_t$ and $Q = q_1 q_2 \cdots q_u$, the p's and q's prime.
From the definition of the Jacobi symbol and part (b) of Proposition (620) we have

$$(P/Q) = \prod_{j=1}^{u} (P/q_j) = \prod_{j=1}^{u} \prod_{i=1}^{t} (p_i/q_j).$$

Let n be the number of i, $1 \leqslant i \leqslant t$, such that $p_i \equiv 3 \pmod 4$, and let m be the
number of j, $1 \leqslant j \leqslant u$, such that $q_j \equiv 3 \pmod 4$. Then the number of ordered
pairs i, j such that $p_i \equiv q_j \equiv 3 \pmod 4$ is mn.

From the Law of Quadratic Reciprocity we see

$$\prod_{j=1}^{u} \prod_{i=1}^{t} (p_i/q_j) = (-1)^{mn} \prod_{j=1}^{u} \prod_{i=1}^{t} (q_j/p_i) = (-1)^{mn} (Q/P).$$

The argument in (671) shows $P \equiv 3 \pmod 4$ if and only if n is odd. In the same way $Q \equiv 3 \pmod 4$ if and only if m is odd. Since $(-1)^{mn} = 1$ unless n and m are both odd, the proposition follows.

(675) **Proposition.** Suppose P, Q_1, and Q_2 are odd positive integers, and $(a,P) = (b,P) = (a,Q_1) = (a,Q_2) = 1$.
  (a) $(a/Q_1 Q_2) = (a/Q_1)(a/Q_2)$
  (b) $(ab/P) = (a/P)(b/P)$
  (c) If $a \equiv b \pmod P$, then $(a/P) = (b/P)$
  (d) $(-1/P) = 1$ if and only if $P \equiv 1 \pmod 4$
  (e) $(2/P) = 1$ if and only if $P \equiv \pm 1 \pmod 8$

(676) **Exercise*.** Prove (675) (a) and (b).

(677) **Exercise*.** Prove (675) (c).

(678) **Exercise*.** Prove (675) (d). [*Hint:* Let $P = p_1 p_2 \cdots p_t$, the p's prime. For how many i is $p_i \equiv -1 \pmod 4$?]

(679) **Exercise*.** Prove (675) (e). [*Hint:* Let $P = p_1 p_2 \cdots p_t$, the p's prime. Suppose $p_i \equiv \pm 3 \pmod 8$ for n values of i. Show n is odd if and only if $P \equiv \pm 3 \pmod 8$.]

(680) Now $(105/211)$ may be calculated as follows:

$$(105/211) = (211/105) = (1/105) = 1.$$

Note that it is not even necessary to check out whether 105 and 211 are prime. In order to draw the conclusion that 105 is a quadratic residue modulo 211, however, it *is* necessary to know that 211 is prime. (It is.)

(681) *Warning.* If Q is not prime, it may be that $(P/Q) = 1$ even though P is not a quadratic residue modulo Q.

(682) **Example.** $(2/15) = (2/3)(2/5) = (-1)(-1) = 1$, yet 2 is not a quadratic residue modulo 15. [How could it be, when $x^2 \equiv 2 \pmod 3$ is not even solvable?]

(683) **Exercise.** Show that if $(P/Q) = -1$, then P is not a quadratic residue modulo Q.

(684) In spite of the above warning, the Jacobi symbol can be useful in the determination of quadratic residues. The symbol $(211/105)$, which appeared in (680), is not a Legendre symbol. Yet the conclusion that 105 is a quadratic residue modulo 211 is perfectly valid.

(685) **Exercise.** Evaluate (77/97), (210/451), (128/315), (361/612), and (-30/79).

(686) **Exercise.** Is 14 a quadratic residue modulo 79? Is 15 one (mod 77)? Is 17 one (mod 77)? Is 5 one (mod 93)?

Extended Exercises

Here follows a series of more substantial exercises. In them the words "prove that" and "show that" have been dropped; each assertion after a letter in parentheses should be proved.

(687) The greatest common divisor and least common multiple of more than two integers.

We define the *greatest common divisor* of the integers $a_1, a_2, \ldots, a_n$ to be the greatest positive integer x such that $x|a_i$, $i = 1, 2, \ldots, n$. It is denoted by $(a_1, a_2, \ldots, a_n)$. We define the *least common multiple* of $a_1, a_2, \ldots, a_n$ to be the smallest positive integer x such that $a_i|x$, $i = 1, 2, \ldots, n$. It is denoted by $[a_1, a_2, \ldots, a_n]$.

(a) If the a's are not all 0, then $(a_1, \ldots, a_n)$ exists. If none of the a's are 0, then $[a_1, \ldots, a_n]$ exists.

(b) If $x|a_i$, $i = 1, 2, \ldots, n$, where the a's are not all 0, then $x|(a_1, \ldots, a_n)$. If $a_i|x$, $i = 1, 2, \ldots, n$, where none of the a's are 0, then
$$[a_1, \ldots, a_n] \,|x.$$

(c) Suppose p is prime and $p^{\alpha_i}||a_i$, $i = 1, 2, \ldots, n$. Then $p^m||(a_1, \ldots, a_n)$ and $p^M||[a_1, \ldots, a_n]$, where $m = \min_{1 \le i \le n} \alpha_i$ and $M = \max_{1 \le i \le n} \alpha_i$.

(d) Suppose $a_1, a_2, \ldots, a_n$ are nonzero integers. Prove or disprove the following statements:
1. $((a_1, a_2), a_3) = (a_1, a_2, a_3)$.
2. $[[a_1, a_2], a_3] = [a_1, a_2, a_3]$.
3. If the a's are relatively prime in pairs, then $(a_1, \ldots, a_n) = 1$.
4. If $(a_1, \ldots, a_n) = 1$, then the a's are relatively prime in pairs.
5. If the a's are relatively prime in pairs, then
$$[a_1, \ldots, a_n] = |a_1, a_2, \ldots, a_n|.$$
6. If $[a_1, \ldots, a_n] = |a_1 a_2 \ldots a_n|$, then the a's are relatively prime in pairs.
7. $(a_1, \ldots, a_n)[a_1, \ldots, a_n] = |a_1 a_2 \ldots a_n|$.

(e) Suppose not all of $a_1, a_2, \ldots, a_n$ are 0. An integer x is a linear combination of $a_1, a_2, \ldots, a_n$ if and only if $(a_1, \ldots, a_n)$ divides x. [*Hint*: Let L be the smallest positive linear combination of the a's.

Show $L = (a_1, \ldots, a_n)$. (See (54).) Another way is to use (49), 1 above, and induction.]

(688) Fermat numbers.

(a) If h and k are positive integers, k odd, then $2^h + 1 | 2^{hk} + 1$.
(b) If $2^n + 1$ is prime, then n is 0 or a power of 2.
The numbers $F_n = 2^{2^n} + 1$ are called *Fermat numbers*. [Compare with (159).] Fermat conjectured they were all prime.
(c) (Euler) F_5 is not prime. [*Hint:* $16 \cdot 2^{28} \equiv -5 \cdot 2^{28} \equiv -(-1)^4 \pmod{641}$.]
Indeed, F_4 is the largest Fermat number known to be prime. F_n has been proved composite for $n = 6, 7, 8, 9, 10, 11, 12, 13$, and certain other values. The case $n = 13$ was only settled in 1960.
(d) If $n \neq m$, then $(F_n, F_m) = 1$. [*Hint:* If $n > m$, then $F_m | F_n - 2$.]
(e) Give an independent proof of (112), based on (d).

(689) Euler's proof that the number of primes is infinite.

(a) Given M there exists n such that $\sum_{i=1}^{n} i^{-1} > M$. [*Hint:* Show $\sum_{i=1}^{2n} i^{-1} \geq (n+1)/2$ by induction on n.]
(b) Suppose S is a set of primes and n is a positive integer such that each prime dividing n is in S. Then there exists k such that if $\prod_{p \in S} (1 + p^{-1} + p^{-2} + \cdots + p^{-k})$ is multiplied out, then $1/n$ is one of the terms.
(c) Give a proof of Theorem (112) based on the above.

(690) The distribution of primes.

The last two exercises not only give alternate proofs of the existence of infinitely many primes, but also provide quantitative information on how thick the primes are. As is standard, we let $\pi(x)$ be the number of primes not exceeding x. For example $\pi(5) = 3 = \pi(\sqrt{40})$. Let p_n be the nth prime, so $p_1 = 2, p_2 = 3$, and $\pi(p_n) = n$.

(a) $\pi(2^{2^n} + 1) \geq n + 2$ for $n \geq 0$. Thus $p_n \leq 2^{2^{n-2}} + 1$ for $n \geq 2$.
(b) Given M there exists k such that $\prod_{i=1}^{k} \left(1 - \frac{1}{p_i}\right)^{-1} > M$. [*Hint:* Use (689) and (125).]
(c) Given any real number $\epsilon > 0$ there exists k such that $\prod_{i=1}^{k} \left(1 - \frac{1}{p_i}\right) < \epsilon$.
(d) For any integer $k \geq 2$, $\prod_{i=2}^{k} \left(1 - \frac{1}{i^2}\right) > \frac{1}{2}$. [*Hint:* Show by induction that $\prod_{i=2}^{k} \left(1 - \frac{1}{i^2}\right) = (k+1)/2k$.]

Since if $\{a_1, a_2, \ldots, a_k\}$ is a set of integers greater than 1 we would expect $\prod_{i=1}^{k} \left(1 - \frac{1}{a_i}\right)$ to be *less* when the a's are small than when they are large, parts (c) and (d) above can be interpreted as indicating that, in some sense, the sequence of primes 2, 3, 5, ... is thicker than the sequence of squares 4, 9, 16, ... Of course (c) may also be expressed by

$$\lim_{k \to \infty} \prod_{i=1}^{k} \left(1 - \frac{1}{p_i}\right) = 0.$$

It can be shown that this implies $\sum_{i=1}^{\infty} p_i^{-1}$ diverges. Thus, in a sense, the primes are thicker than any sequence of integers the sum of the reciprocals of which converges.

(691)   Counting solutions of $\varphi(x) = 2^n$.

Let $N(n)$, $E(n)$, and $O(n)$ be, respectively, the number of solutions, the number of even solutions, and the number of odd solutions x of the above equation.

(a) $N(0) = 2, N(1) = 3, N(2) = 4, N(3) = 5, O(0) = O(1) = O(2) = O(3) = 1$.

It can be calculated that $N(n) = n + 2$ for $0 \leqslant n \leqslant 31$.

(b) If $n \geqslant 1$, then $N(n) = E(n-1) + 2 \cdot O(n)$.

(c) $N(n) = n + 2$ for all $n \geqslant 0$ if and only if $O(n) = 1$ for all $n \geqslant 0$.

(d) The odd number $x > 1$ is a solution of $\varphi(x) = 2^n$ if and only if x is the product of distinct Fermat primes, i.e., prime Fermat numbers.

(e) If n is any nonnegative integer there exists a unique sequence of numbers $a_0, a_1, a_2, \ldots$ such that each a_i is 0 or 1 and $n = \sum_{k=0}^{\infty} a_k 2^k$. [*Hint*: Show that given t there are 2^t such sequences with $a_i = 0$ for all $i \geqslant t$. For each such sequence $0 \leqslant \sum_{k=0}^{\infty} a_k 2^k < 2^t$. Thus it suffices to show these sums are distinct. Suppose

$$\sum_{k=0}^{t-1} a_k 2^k = \sum_{k=0}^{t-1} b_k 2^k.$$

Show $a_0 = b_0, a_1 = b_1$, etc.]

(f) $O(n) \leqslant 1$ for $n \geqslant 0$ and $O(n) = 1$ for $0 \leqslant n \leqslant 31$. $O(32) = 0$. $N(n) = n + 2$ for $0 \leqslant n \leqslant 31$, while $N(n) = 32$ for $32 \leqslant n \leqslant 16383$. [You may assume the remark following (688) (c).]

(692)   The equation $x^2 + y^2 = z^2$.

Above is probably the most famous example of a *Diophantine equation*, i.e., an equation (usually involving two or more unknowns) for which a solution is desired in integers (or, sometimes, rational numbers). The name comes from the Greek mathematician Diophantus, who probably lived during the third century A.D.

Since if $k \neq 0$, then x, y, z is a solution of $x^2 + y^2 = z^2$ if and only if kx, ky, kz is, we will restrict our attention to *primitive* solutions, that is, solutions with $(x,y,z) = 1$. We will consider only positive x,y, and z.

(a) If x, y, z is a primitive solution of $x^2 + y^2 = z^2$, then x, y, and z are relatively prime in pairs.

(b) If x, y, z is a primitive solution of $x^2 + y^2 = z^2$, then x and y are neither both even nor both odd. [*Hint*: $x^2 + y^2 \equiv ?$ (mod 4).]

(c) Suppose x, y, z is a primitive solution of $x^2 + y^2 = z^2$ with x even. Then $(\frac{1}{2}(z - y), \frac{1}{2}(z + y)) = 1$.

(d) If x, y, z is a primitive solution of $x^2 + y^2 = z^2$ with x even, then there exist integers r and s, not both odd, with $(r, s) = 1$ and $0 < r < s$, such that $x = 2rs$, $y = s^2 - r^2$, and $z = r^2 + s^2$.

(e) Suppose $0 < r < s$, $(r, s) = 1$, and r and s are not both odd. Set $x = 2rs$, $y = s^2 - r^2$, and $z = r^2 + s^2$. Then x, y, z is a primitive solution of $x^2 + y^2 = z^2$.

(f) Find all solutions of $x^2 + y^2 = z^2$ with $0 < x \leqslant y \leqslant z \leqslant 25$. (There are eight.)

Triples of positive integers x, y, z satisfying $x^2 + y^2 = z^2$ are called *Pythagorean triples*, since they represent the sides of right triangles.

(693) Fermat's Last Theorem.

The theorem referred to says that if n is any integer greater than 2 the equation $x^n + y^n = z^n$ has no solutions in positive integers. Pierre de Fermat, a French lawyer, was the best mathematician of his day, despite regarding mathematics only as a hobby. He was supposedly born in 1601 and died in 1665—at the age (according to the wording on his tombstone) of 57. Life was slower then.

Fermat stated his theorem in the margin of his copy of Diophantus's works, adding "I have discovered a truly remarkable proof which this margin is too small to contain." He never gave his proof, nor has anyone given a proof since. Neither has a counterexample been found. It is the most famous unsolved problem of mathematics.

(a) To prove Fermat's Last Theorem it suffices to show

(*) $x^n + y^n = z^n$

is impossible in positive integers if $n = 4$ and if n is any odd prime.

If $n = 4$ we see (*) is a special case of the equation of (692). So is

(**) $x^4 + y^4 = z^2$.

It turns out to be more convenient to treat the latter equation, somewhat more general than $x^4 + y^4 = z^4$. As before, x, y, z is a *primitive* solution of (**) if $(x, y, z) = 1$.

(b) If x, y, z is a solution of (**), then x and y are not both odd.

In parts (c), (d), and (e) below we assume x, y, z is a primitive solution of (**) in positive integers with x even.

(c) There exist integers r and s such that $0 < r < s$, $(r, s) = 1$, $2 | r$, $2 \nmid s$, $x^2 = 2rs$, $y^2 = s^2 - r^2$, and $z = r^2 + s^2$.

(d) With r and s as above, there exist integers R and S such that $r = 2RS$, $y = S^2 - R^2$, $s = R^2 + S^2$, with $(R, S) = 1$ and $0 < R < S$. [*Hint*: Apply (692) to $r^2 + y^2 = s^2$.]

(e) With r, s, R, and S as above, there exist positive integers X, Y, and Z such that $X^2 = R$, $Y^2 = S$, and $Z^2 = s$. [*Hint*: Start by showing $r/2$ and s are squares.]

(f) If x, y, z is any primitive solution of (**) in positive integers there exists another positive primitive solution X, Y, Z with $Z < z$.

(g) There is no solution of $x^4 + y^4 = z^2$ in positive integers. [*Hint*: We have seen any such solution generates another solution X, Y, Z with $0 < Z < z$. How long can this go on?]

The argument used in (g), called the "method of infinite descent," is ascribed to Fermat. It is a somewhat ingenious relative of mathematical induction.

(h) Fermat's Last Theorem is true for $n = 4$.

(694) Rational and irrational numbers.

We say the real number x is *rational* if $x = a/b$, where a and b are integers, $b \neq 0$. Otherwise we call x *irrational*. Most everyone has seen a proof that $\sqrt{2}$ is irrational, a fact discovered by the Greeks. (They didn't like it.)

(a) If k is not the nth power of some integer, then $\sqrt[n]{k}$ is irrational.

(b) Suppose r and r' are rational numbers, $r' \neq 0$, and suppose s and s' are irrational numbers. Prove or disprove:

1. $r + r'$ is rational.
2. rr' is rational.
3. $s + s'$ is irrational.
4. ss' is irrational.
5. $r + s$ is irrational.
6. rs is irrational.
7. $r's'$ is irrational.

(c) Suppose a and b are real numbers, $a < b$. There exists a rational number r and an irrational number s such that $a < r < b$ and $a < s < b$.

The rest of this exercise requires an acquaintance with infinite series. We define e to be $\sum_{i=0}^{\infty} \frac{1}{i!}$, where $0! = 1$.

(d) If $k \geqslant 1$, then

$$\frac{1}{k+1} + \frac{1}{(k+1)(k+2)} + \frac{1}{(k+1)(k+2)(k+3)} + \cdots < 1.$$

(e) If k is an integer $\geqslant 1$, then $k!e$ is not an integer.

(f) The number e is irrational.

(695) Primitive polynomials; Gauss's Lemma.

We call α a *zero* of the polynomial $P(x)$ (which may or may not be integral) in case $P(\alpha) = 0$. We call $P(x)$ *monic* in case the coefficient of the highest power of x is 1.

(a) Suppose the nonzero rational number h/k, $(h,k) = 1$, is a zero of the (integral) polynomial $a_n x^n + a_{n-1} x^{n-1} + \cdots + a_0$. Then $k|a_n$ and $h|a_0$.

(b) All the rational zeros of any monic (integral) polynomial are integers. An integral polynomial $a_n x^n + \cdots + a_0$ is said to be *primitive* in case $(a_n, a_{n-1}, \ldots, a_0) = 1$.

(c) If $f(x)$ is any polynomial with rational coefficients, then there exists a rational number $a/b = r$ such that $r \cdot f(x)$ is a primitive integral polynomial. If f is monic we may take $b = 1$, while if f is already an integral polynomial we may take $a = 1$.

(d) The product of primitive polynomials is primitive. [*Hint*: Suppose the prime p divides each coefficient of $f(x)g(x)$, where $f(x) = a_n x^n + \cdots + a_0$ and $g(x) = b_m x^m + \cdots + b_0$ are primitive. Let i be minimal such that $p \nmid a_i$; let j be minimal such that $p \nmid b_j$. Show p does not divide the coefficient of x^{i+j} in $f(x)g(x)$.]

Part (b) says that if a monic integral polynomial has a monic linear factor with rational coefficients, these coefficients (there is really only one besides 1) are in fact integers. The following is a generalization.

(e) Gauss's Lemma. If the monic integral polynomial f can be written as $f_1 f_2$, where f_1 and f_2 are also monic and have rational coefficients, then the coefficients of f_1 and f_2 are in fact integers. [*Hint*: Find integers a_1 and a_2 such that $a_1 f_1$ and $a_2 f_2$ are primitive. Then $a_1 a_2 f_1 f_2 = a_1 a_2 f$ is primitive. Conclude $a_1 a_2 = \pm 1$.]

(696) Approximation by rationals.

Part (c) of (694) shows that any real number has rational numbers arbitrarily close to it. Still some rational approximations are "better" than others. For example $22/7$ is considered to be a "good" approximation to π. Why is it better than, say, $3.143 = 3143/1000$, which is about as close? It is simpler. A good way to measure the simplicity of a fraction is by the smallness of its denominator. (The numerator must depend on the size of the number; for example, any rational approximation to a number exceeding 1,000,000 must have a numerator at least 1,000,000, even though the denominator may be small.) Thus a good approximation: $1°$ is close, and $2°$ has a small denominator.

(a) Given a real number α and a positive integer k there exists an integer h such that $|\alpha - h/k| < 1/k$.

More is true. Making $|\alpha - h/k|$ small is like making $|k\alpha - h|$ small. If α is any real number we define $[\alpha]$ to be the greatest integer not exceeding α. Thus $[\pi] = 3$, $[-\pi] = -4$, and $[5] = 5$.

(b) Let n be a positive integer. Suppose α_k is a real number with $0 \leqslant \alpha_k < 1$ for $k = 0, 1, \ldots, n$. Then there exist i and j, $0 \leqslant i < j \leqslant n$, such that
$$|\alpha_i - \alpha_j| < 1/n.$$

(c) If α is any real number there exist i and j, $0 \leqslant i \leqslant n$, such that
$$|(i\alpha - [i\alpha]) - (j\alpha - [j\alpha])| < 1/n.$$

(d) If n is a positive integer and α is any real number there exist integers h and k, $0 < k \leqslant n$, such that $|k\alpha - h| < 1/n$.

(e) If α is any real number there exists a rational number h/k such that $|\alpha - h/k| < 1/k^2$.

(f) (Dirichlet) If α is any irrational number there exist infinitely many rational numbers h/k, $(h,k) = 1$, such that $|\alpha - h/k| < 1/k^2$. [*Hint*: Suppose only h_1/k_1, h_2/k_2, $\ldots$, h_t/k_t satisfy the inequality. Choose n such that $1/n < \min_{1 \leqslant i \leqslant t} |\alpha - h_i/k_i|$. Apply (d).]

(697) Order of approximation.

We say the real number α is *approximable to order m* if there exists a positive number C (which may depend on α) such that there exist infinitely many rational numbers h/k with $(h,k) = 1$ such that
$$|\alpha - h/k| < C/k^m.$$

The last exercise shows each irrational number is approximable to order 2. (Take $C = 1$.)

(a) Suppose C and ϵ are positive real numbers. Consider the rational number a/b, $b > 0$. There exist only finitely many rationals h/k with $(h,k) = 1$ and $k > 0$ such that
$$\left| \frac{a}{b} - \frac{h}{k} \right| < \frac{C}{k^{1+\epsilon}}.$$

[*Hint*: If $a/b \neq h/k$, then $|a/b - h/k| \geqslant 1/bk$. But $1/bk \leqslant C/k^{1+\epsilon}$ for only finitely many k.]

(b) Any rational number is approximable to order 1 but to no higher order.

We say the number α is *algebraic* if it is a zero of a polynomial with integral coefficients. We say α is *algebraic of degree m* if α is a zero of some (integral) polynomial of degree m but not of any (integral) polynomial of lower degree. A real number that is not algebraic is said to be *transcendental*.

(c) No algebraic number of degree 1 is approximable to any order greater than 1.

(d) Let $\alpha_0 = \Sigma_{n=1}^{\infty} 10^{-n!}$. Show that if m is any positive real number, then α_0 is approximable to order m. [*Hint:* Let $r_N = \Sigma_{n=1}^{N} 10^{-n!}$. Show that if $r_N = h_N/k_N$, with $(h_N, k_N) = 1$ and $k_N > 0$, then $k_N = 10^{-N!}$. Show that if $N \geqslant m$, then $|\alpha_0 - r_N| < 2/k_N^{N+1} < 2/k_N^m$.]

(698) Liouville's Theorem.

Liouville's Theorem says that no algebraic number of degree m is approximable to any order greater then m. The case $m = 1$ is covered by part (c) of the last exercise.

We assume throughout this exercise that α is an algebraic number of degree m, $m > 1$, and that α is a zero of the integral polynomial

$$P(x) = a_m x^m + \cdots + a_1 x + a_0.$$

Let ϵ be a positive real number.

(a) If C is any positive real number, then there exist at most a finite number of rationals h/k, $(h,k) = 1$, such that $|h/k - \alpha| \geqslant 1$ and $|\alpha - h/k| \leqslant C/k^m$.

(b) If $P(h/k) \neq 0$, then $|P(h/k)| \geqslant 1/k^m$.

(c)

$$-P(r) = (\alpha - r)\{a_m(\alpha^{m-1} + \alpha^{m-2}r + \cdots + r^{m-1})$$
$$+ a_{m-1}(\alpha^{m-2} + \alpha^{m-3}r + \cdots + r^{m-2})$$
$$+ \cdots$$
$$+ a_1\}.$$

[*Hint:* $-P(r) = P(\alpha) - P(r) = a_m(\alpha^m - r^m) + \cdots + a_1(\alpha - r)$. Use (125).]

(d) There exists a positive number K such that if $|\alpha - r| < 1$, and $\alpha \neq r$, then $|-P(r)/(\alpha - r)| \leqslant K$. [*Hint:* Let $t = |\alpha| + 1$. Take $K = |a_m| \cdot (|\alpha|^{m-1} + |\alpha|^{m-2}t + \cdots + t^{m-1}) + |a_{m-1}|(|\alpha|^{m-2} + \cdots + t^{m-2}) + \cdots + |a_1|.$]

(e) There exist only finitely many rationals $h/k = r$, with $(h,k) = 1$, such that $|\alpha - h/k| < 1/Kk^m$.

(f) If C is any positive number, then there exist only finitely many rationals $r = h/k$, with $(h,k) = 1$, such that $|\alpha - h/k| < C/k^{m+\epsilon}$.

(g) The number α_0 defined in the previous exercise is transcendental.

(699) Farey fractions.

Let n be a positive integer. The rational numbers h/k with $0 \leqslant h \leqslant k \leqslant n$, $(h,k) = 1$, and $k \neq 0$, listed in increasing order, are called the *Farey sequence of order n*. The Farey sequence of order 4, for example, is

$$0 = \frac{0}{1}, \frac{1}{4}, \frac{1}{3}, \frac{1}{2}, \frac{2}{3}, \frac{3}{4}, \frac{1}{1} = 1.$$

It is easy enough to determine the elements of the Farey sequence of order n, but putting them in increasing order may be troublesome. The reader may like to order the numbers 0, 1, $\frac{1}{2}$, $\frac{1}{3}$, $\frac{2}{3}$, $\frac{1}{4}$, $\frac{3}{4}$, $\frac{1}{5}$, $\frac{2}{5}$, $\frac{3}{5}$, $\frac{4}{5}$, $\frac{1}{6}$, and $\frac{5}{6}$, which are the elements of the Farey sequence of order 6.

(a) If k and k' are positive and $h/k < h'/k'$, then

$$\frac{h}{k} < \frac{h+h'}{k+k'} < \frac{h'}{k'}.$$

Part (a) can help with the ordering problem. Let us start with the Farey sequence of order 1, namely,

$$\frac{0}{1}, \frac{1}{1}.$$

We now insert the number obtained by adding numerators and denominators, $\frac{0+1}{1+1} = \frac{1}{2}$. Now we have

$$\frac{0}{1}, \frac{1}{2}, \frac{1}{1}.$$

Call the above sequence step 2; it happens to be the Farey sequence of order 2. We now treat the adjacent fractions in exactly the same way, again adding numerators and denominators. At step 3 we have

$$\frac{0}{1}, \frac{1}{3}, \frac{1}{2}, \frac{2}{3}, \frac{1}{1}.$$

Of course (a) assures us that the sequence we obtain at each step this way is in increasing order. At step 3 we got the Farey sequence of order 3.

We proceed to step 4:

$$\frac{0}{1}, \frac{1}{4}, \frac{1}{3}, \frac{2}{5}, \frac{1}{2}, \frac{3}{5}, \frac{2}{3}, \frac{3}{4}, \frac{1}{1}.$$

This is *not* the Farey sequence of order 4. (It contains $\frac{2}{5}$ and $\frac{3}{5}$.) Neither is it the Farey sequence of order 5. (Both $\frac{1}{5}$ and $\frac{4}{5}$ are missing.) It does *contain* the Farey sequence of order 4, however. Let us modify our procedure in light of the above. Let us agree to insert $\frac{h+h'}{k+k'}$, between h/k and h'/k' at the nth step only if $k + k' \leqslant n$.

With this modification we have at the fourth step

$$\frac{0}{1}, \frac{1}{4}, \frac{1}{3}, \frac{1}{2}, \frac{2}{3}, \frac{3}{4}, \frac{1}{1}.$$

This is the Farey sequence of order 4. At the 5th step we refrain from inserting anything between $\frac{1}{4}$ and $\frac{1}{3}$ or between $\frac{2}{3}$ and $\frac{3}{4}$. Thus we get

$$\frac{0}{1}, \frac{1}{5}, \frac{1}{4}, \frac{1}{3}, \frac{2}{5}, \frac{1}{2}, \frac{3}{5}, \frac{2}{3}, \frac{3}{4}, \frac{4}{5}, \frac{1}{1},$$

which is the Farey sequence of order 5.

(b) If h/k and h'/k' are adjacent fractions at some step in the procedure described above, with $h/k < h'/k'$, then $h'k - hk' = 1$. [*Hint*: This is true of the sequence $\frac{0}{1}, \frac{1}{1}$. Use induction.]

(c) If h/k appears at any step, then $(h,k) = 1$.

(d) Suppose h/k and h'/k' are adjacent fractions at some step. (Either may be larger.) Then there is no fraction with denominator $\leqslant k$ strictly between them. [*Hint*: Suppose a/b is. Show $|h/k - h'/k'| = 1/kk'$, while $|h'/k' - a/b| \geqslant 1/k'b$.]

(e) The nth step in the procedure outlined above gives the Farey sequence of order n. [*Hint*: Suppose m/n doesn't appear at the nth step, where $0 < m/n < 1$ and $(m,n) = 1$. Say $h/k < m/n < h'/k'$, where h/k and h'/k' are adjacent fractions at the nth step. Prove $k + k' > n$. At the $(k + k')$th step we will insert $\dfrac{h + h'}{k + k'}$, while m/n will never be inserted. Show that this contradicts (d).]

(f) If h/k, h'/k', and h''/k'' are three consecutive terms of the Farey sequence of order n, in increasing order, then $h'k - hk' = 1$ and $\dfrac{h'}{k'} = \dfrac{h + h''}{k + k''}$. [*Hint*: The first equation follows from (b) and (e). Solving the equations $h'k - hk' = 1$ and $h''k' - h'k'' = 1$ for h' and k' gives the second.]

Some Books to Look at

Although all elementary number theory books cover more or less the same ground, the arrangement of the material and the proofs given vary widely. Thus reading another treatment of a topic will usually add to one's understanding of it. Any number theory book is good this way, and the ones listed below are only a selection of those available.

Three texts at about the same level as this one (but more conventional in style) are William J. LeVeque, *Elementary Theory of Numbers* (Addison-Wesley, 1962), Neal H. McCoy, *The Theory of Numbers* (Macmillan, 1965), and I. A. Barnett, *Elements of Number Theory* (Prindle, Weber and Schmidt, 1968).

At a somewhat higher level are LeVeque, *Topics in Number Theory*, 2 vols. (Addison-Wesley, 1956), and Ivan Niven and H. S. Zuckerman, *An Introduction to the Theory of Numbers*, 2d ed. (Wiley, 1966). These texts contain more material and more sophisticated proofs, and are often used for first-year graduate courses.

In a class by itself is G. H. Hardy and E. M. Wright, *An Introduction to the Theory of Numbers*, 4th ed. (Oxford, 1960). This is not a textbook (there are no exercises, for example), but a full and engagingly written account of the subject which provides pleasant browsing to enthusiasts of all degrees of preparation.

A trip to the library to examine L. E. Dickson, *History of the Theory of Numbers*, 3 vols., is well justified. The original version published by the Carnegie Institute of Washington (1919, 1920, 1923) will probably be there; a reprint by the Chelsea Publishing Company is now available. Dickson took upon himself the job of summarizing all the papers in number theory written up until his work was published; and anyone interested in the sources of the theorems in this book will find them, names and dates, in his *History*.

Number theory is lucky to have had its origins investigated thoroughly; Oystein Ore, *Number Theory and Its History* is an interesting account of them.

Some paperback books are L. E. Dickson, *Introduction to the Theory of Numbers* (Dover Publications, reprint of 1929 ed.), Robert D. Carmichael, *The Theory of Numbers* and *Diophantine Analysis* (Dover, reprint of 1914 and 1915 eds.), I. M. Vinogradov, *Elements of Number Theory*, 5th rev. ed. (Dover, 1954), Harriet Griffin, *Elementary Theory of Numbers* (McGraw-Hill), and H. Davenport, *The Higher Arithmetic: An Introduction to the Theory of Numbers* (Harper, 1960, also Hutchinson University Library, 1968). The first two listed are reprints of fairly old works, but the subject is old. The book by Vinogradov, a Russian mathematician, is mainly problems, and anyone working through them all will wind up with a respectable technique in analytic number theory. Griffin's is a textbook, also available between hard covers. Davenport gives a delightful modern introduction all the way to the frontiers of the subject.

A field untouched in this book is *algebraic number theory*, which, roughly, treats numbers and other things that act like integers, even if they are not. Such investigations often yield results about the ordinary integers. A good introduction is Harry Pollard, *The Theory of Algebraic Numbers* (Carus Mathematical Monographs, No. 9, Wiley, 1950).

Answers to Numerical Exercises

(30)

a	b	$[a,b]$	ab	(a,b)	$ab/(a,b)$
6	-12	12	-72	6	-12
7	-12	84	-84	1	-84
8	-12	24	-96	4	-24
9	-12	36	-108	3	-36
10	-12	60	-120	2	$-60.$

(40) $q = 29, r = 3; q = 0, r = 13; q = -30, r = 10.$

(52) $(a,b) = 7, x = 2, y = -1; (a,b) = 1, x = -377, y = 610; (a,b) = 1, x = -446, y = 45.$

(62)

b	$(4,b)$	$(5,b)$	b	$(4,b)$	$(5,b)$
1	1	1	11	1	1
2	2	1	12	4	1
3	1	1	13	1	1
4	4	1	14	2	1
5	1	5	15	1	5
6	2	1	16	4	1
7	1	1	17	1	1
8	4	1	18	2	1
9	1	1	19	1	1
10	2	5	20	4	5.

(108) $293^1, 7^1 11^1 13^1, 41^1 43^1, 2^1 3^1 5^1 7^1 11^1, 2^3 3^6, 2^6 5^6.$

(110) 1009, 1013, 1019, 1021.

(112)

n	(P,n)	n	(P,n)
11	11	16	2
12	6	17	17
13	13	18	6
14	14	19	19
15	15	20	10.

(116) The divisors of 360 are 1, 2, 3, 4, 5, 6, 8, 9, 10, 12, 15, 18, 20, 24, 30, 36, 40, 45, 60, 72, 90, 120, 180, 360. $\tau(210) = 16, \tau(243) = 6, \tau(244) = 10, \tau(1000000) = 49.$

(126) $\sigma(40) = 90, \sigma(10^6) = 2480437, \sigma(360) = 1248.$

(129) 50, 91, 27, 273.

(136) $0, -24, 2, 2, 0, 0, 14400, 36, 0, 1.$

(146)

n	$\sigma(n)$	n	$\sigma(n)$
31	32	36	91
32	63	37	38
33	48	38	60
34	54	39	56
35	48	40	90.

(150) 8128.

(164) No, since $2^{49} = (2^3)^{16} 2 \equiv 1^{16} 2 \pmod 7$.

(173) $f(p^2 q^2) = F(p^2 q^2) - F(p^2 q) - F(pq^2) + F(pq)$

(182) $1, 0, -1, 0, -1, 0, 1.$

(184) d.

(206) $2, 6, 18, 8.$

(209) $12, 36, 64, 162.$

(212) $72, 72, 720.$

(216) $16, 16, 16, 16, 16, 16.$

(218) $n_0 = 101^8$ works. (Suppose $n \geqslant 101^8$. If some prime $\geqslant 101$ divides n, we are done. Otherwise n must have at least 8 prime divisors (not necessarily distinct). Thus $\varphi(n) \geqslant \varphi(2^8) = 128$.)

(220) $\varphi(1) = \varphi(2) = 1, \varphi(3) = \varphi(4) = \varphi(6) = 2, \varphi(5) = \varphi(8) = \varphi(10) = \varphi(12) = 4.$

(221) No solution.

(229) $n = 1, 2.$

(246) $n = 1, 11; m = 1, 13, 7, 19; nm = 1, 7, 11, 13, 19, 77, 143, 209.$

(253) One is $9, 27, 45, 63.$

(254) $x = 6.$

(255) $x = 24, x = -4.$

(256) (a) $x = 4$; (b) $x = 2, 5$; (c) No solution.

(258) One is $x = 7n$.

(267) Any $x \equiv 8 \pmod{30}$.

(268) Any $x \equiv 53 \pmod{60}$.

(270)

(n,m)	k
$(1,5)$	11
$(-1,5)$	-1
$(2,5)$	-13
$(8,5)$	-7
$(1,7)$	1
$(-1,7)$	-11
$(2,7)$	7
$(8,7)$	$13.$

(271) In the first case $n = 1, 3$ and $m = 1, 3, 7, 9$. Then $1 \cdot 3 \equiv 3 \cdot 1 \pmod{20}$. In the second case $nm = 1, 3, 7, 9, 11, 33, 77, 99.$

(277) $4t + 5s = 9, 13, 17, 19, 21, 23, 27, 31.$
$4u + 5t = 9, 13, 14, 17, 18, 19, 21, 22, 23, 24, 25, 26, 27, 28, 30, 31, 32, 35, 36, 40.$

(283) (a) $k = -22$; (b) $k = 19$; (c) $k = -10,099$; (d) $k = 53$.
(288) $x = 59$.
(289) $x = 51$.
(290) $x = -65$.
(291) No solution.
(294) $\xi = 3$.
(296) (a) $x = 6, 2, 4$; (b) $x = 3$; (c) $x = 10, 4, -7$.
(297) $\xi = 31, 51, 52, 72, 121, 142, 171, 192$.
(298) $\xi = 23, 33, 43, 68, 78, 88$.
(301) Any $\xi \equiv 23 \pmod{105}$.
(303) $\alpha = 21$.
(304) $\theta = -418$.
(305) $k = 77, 197$.
(306) 2000.
(307) 1986.
(312) (a) $x = 11$; (b) $x = 266$; (c) $x = 100, 19$.
(313) $\xi = 69,997$.
(314) $x = 1493$.
(322) (a) One is $x = 14$.
       (b) One is $y = -2, 8, 18$.
       (c) One is $z = 68$.
(330) One is $x = 28, 121, 214, 307, 400, 493$.
(331) One is $x = 25 + 53k, k = 0, 1, \ldots, 99$.
(332) No solution.
(333) No solution.
(337) One is $x = 0, 1, 15, 21$.
(338) One is $x = 0, 12, 15, 27, 30, 42$.
(339) One is $x = 1, 5, 7, 11, 13, 17, 19, 23$.
(341) There are 1, 2, and 6 elements, respectively. It had better not!
(346) One is $x = \pm 1, \pm 7, \pm 11, \pm 13$.
(347) One is $x = -1, 17, 26$.
(348) One is $x = -5$.
(349) One is $x = 37, 47, 107, 117, 177, 187$.
(351) (a) One is $x = \pm 12$.
       (b) One is $x = -1, 4, 9, 14, 19$.
       (c) One is $x = -2, 3$.
       (d) One is $x = 0, \pm 1, \pm 7$.
(354) $x = 44$. Yes.
(367) (a) $x = 97$; (b) $y = 76$; (c) $z = 21$.
(379) $x = -1, 0, \pm 5, \pm 10$.
(380) $x = -2$.
(381) $x = \pm 1, \pm 7$.
(382) $x = \pm 1$.
(383) $x = \pm 38$.

(387) $x = -4$.
(388) No solution.
(389) $x = -25, 22$.
(390) $x = 5$.
(391) $x = 4$.
(397) $h = 0, 1, 2, 3, 4, 5, 6, 7, 9$.
(398) $x = -3, 4, 372, 629$.
(399) $x = 1$.
(400) $x = 5, 14$.
(401) No solution.
(409) $x = -3, 11$.
(410) $x = 1, 3$.
(411) No solution.
(412) $x = 55, 56$.
(413) No solution.
(414) $x = 9, 43$.
(420) $a = 1, 4, 5, 6, 7, 9, 11, 16, 17$.
(421) $a = 1, 9$.
(424) $x = 9, 23, 41, 55$.
(434) $1, 4, 4, 2; 1, 10, 5, 5, 5, 10, 5, 10, 5, 2$.
(439) $1, 12, 3, 6, 4, 12, 12, 4, 3, 6, 12, 2$. Yes.
(443) $x = 13$.
(444) $x = 19$.
(445) $x = 25$.
(446) $x = \pm 3$.
(447) $x = 0$.
(453) $x = 3, 7$.
(454) No solution.
(455) No solution.
(459) $2 \cdot 7 \equiv 3 \cdot 9 \equiv 4 \cdot 10 \equiv 5 \cdot 8 \equiv 6 \cdot 11 \equiv 1 \pmod{13}$.
(463) $x = 1$.
(464) $a = 1, 3, 4, 5$.
(465) No.
(466) Yes.
(467) No.
(468) No.
(471) No.
(472) No.
(473) No.
(474) Yes.
(479) (a) 3; (b) 2; (c) 1; (d) No degree; (e) 0.
(485) (a) $q(x) = 2, r(x) = -3x + 2$.
       (b) $q(x) = 0, r(x) = 2x^2 + 3x + 4$.
       (c) $q(x) = x^3 + x^2 + x + 1, r(x) = 0$.

(493) 96.
(502) 1, 1, and 2 elements of orders 1, 2, and 4, respectively.
(513) 1, 1, 2, 2, 2, and 4 elements of orders 1, 2, 3, 4, 6, and 12, respectively.
(514) 1, 1, 2, 4, and 8 elements of orders 1, 2, 4, 8, and 16, respectively.
(515) 1, 1, 2, and 2 elements of orders 1, 2, 3, and 6, respectively.
(516) 1, 3, and 4 elements of orders 1, 2, and 4, respectively.
(521) 3, 7; 2, 6, 7, 8; none; 2, 6, 7, 11.
(522) 7, 13, 17, 19.
(528) $A = 3, 27, 5, 11, 31, 7, 29, 23.$
(529) $A = 2, 8, 3, 12, 23, 17, 22, 13.$
(530) None exist.
(531) 2.
(532) No. No.
(533) Yes.
(536) 61 and 62.
(541) Do: 94, 97, 101, 103; do not: 95, 96, 99, 100, 102.
(547)

m	Primitive root?	Reason
11	Yes	Proposition (524)
12	No	Proposition (540)
13	Yes	Proposition (524)
14	Yes	Propositions (524) and (526)
15	No	Proposition (540)
16	No	Proposition (545)
17	Yes	Proposition (524)
18	Yes	Exercise (514)
19	Yes	Proposition (524)
20	No	Proposition (540)
21	No	Proposition (540)
22	Yes	Propositions (524) and (526)
23	Yes	Proposition (524)
24	No	Proposition (540)
25	Yes	Exercise (529).

(549) $A = 2, A = 3.$
(553) ③ ⑩ ⑰ ㉔ 31 ㊳ ㊺
       ⑤ ⑫ 19 ㉖ ㉝ ㊵ ㊼.
(566) $m = 2, m = 3,$ none.
(569) 8, 14, 30.
(575) $.\overline{3}, .\overline{1}, .\overline{076923}$; 1, 1, 6.
(581) $.\overline{0588235294117647}.$
(582) 18.
(583) 6.
(584) 58.
(587) $m = 47.$
(589) $m = 11, 33, 99.$
(590) $m = 27, 37, 111, 333, 999.$

(600)

0	1	2	3	4	5	6	7	8	9	10	11
1	15	17	21	3	19	25	11	9	5	23	7;

5, 8.

(601) (a) 1; (b) 17; (c) 23; (d) 15; (e) 11.

(605) 1, 17, 3, 25, 9, 23.

(606) 1, 5, 21, 25; indices 0, 9, 3, 6.

(607) 0, 3, 6, 9, 12, 15, 18, 21, 24, 27, 30, 33, 36, 39.

(608) 0, 1, 2, 3, 4, 5, 6, 7, 8, 9, 10.

(615) No.

(618) 1, 1, undefined, 1, undefined, undefined, -1, undefined, -1.

(623)

s	1	2	3	4	5
s^*	3	-5	-2	1	4
$|s^*|$	3	5	2	1	4;

$(3/11) = 1$.

(624)

s	1	2	3	4	5	6
s^*	5	-3	2	-6	-1	4
$|s^*|$	5	3	2	6	1	4;

$(5/13) = -1$.

(630)

s	1	2	3	4	5	6	7	8	9	10	11
s^*	7	-9	-2	5	-11	-4	3	10	-6	1	8;

$n = 5, (7/23) = -1$.

(631)

s	1	2	3	4	5	6	7	8	9	10	11
s^*	2	4	6	8	10	-11	-9	-7	-5	-3	-1;

$n = 6, (2/23) = 1$.

(632) $n = 25, (2/101) = -1$

(639) $-1, 1, -1, -1$.

(640) $-1, -1, 1, -1$.

(641) $(p-1)/4, (p+1)/4, (p-1)/4, (p+1)/4$.

(642) $(p+5)/4$.

(644) No, no, no, yes.

(649) $-1, 1, -1, 1, 1$.

(653) $-1, -1, 1$.

(658) $1, -1, -1$.

(662) $1, -1, 1, -1$.

(667) $-1, -1, -1, 1$.

(668) No, yes, no, no.

(685) $-1, 1, -1$, undefined, -1.

(686) No, yes, no, no.

Index